393 Lessons from Nature!

R/F!

Evolution, Old & New

Or the Theories of Buffon, Dr. Erasmus Darwin and Lamarck, as compared with that of Charles Darwin

by

Samuel Butler

"The want of a practical acquaintance with Natural History leads the author to take an erroneous view of the bearing of his own theories on those of Mr. Darwin.--*Review of 'Life and Habit,' by Mr. A. R. Wallace, in 'Nature,' March 27, 1879.*

Originally published in 1911.

"Neither lastly would our observer be driven out of his conclusion, or from his confidence in its truth, by being told that he knows nothing at all about the matter. He knows enough for his argument; he knows the utility of the end; he knows the subserviency and adaptation of the means to the end. These points being known, his ignorance concerning other points, his doubts concerning other points, affect not the certainty of his reasoning. The consciousness of knowing little need not beget a distrust of that which he does know."

Paley's '*Natural Theology*,' chap. i.

NOTE

The demand for a new edition of "Evolution, Old and New," gives me an opportunity of publishing Butler's latest revision of his work. The second edition of "Evolution, Old and New," which was published in 1882 and re-issued with a new title-page in 1890, was merely a re-issue of the first edition with a new preface, an appendix, and an index. At a later date, though I cannot say precisely when, Butler revised the text of the book in view of a future edition. The corrections that he made are mainly verbal and do not, I think, affect the argument to any considerable extent. Butler, however, attached sufficient importance to them to incur the expense of having the stereos of more than fifty pages cancelled and new stereos substituted. I have also added a few entries to the index, which are taken from a copy of the book, now in my possession, in which Butler made a few manuscript notes.

R. A. STREATFEILD.

October, 1911.

AUTHOR'S PREFACE

TO

THE SECOND EDITION

Since the proof-sheets of the Appendix to this book left my hands, finally corrected, and too late for me to be able to recast the first of the two chapters that compose it, I hear, with the most profound regret, of the death of Mr. Charles Darwin.

It being still possible for me to refer to this event in a preface, I hasten to say how much it grates upon me to appear to renew my attack upon Mr. Darwin under the present circumstances.

I have insisted in each of my three books on Evolution upon the immensity of the service which Mr. Darwin rendered to that transcendently important theory. In "Life and Habit," I said: "To the end of time, if the question be asked, 'Who taught people to believe in Evolution?' the answer must be that it was Mr. Darwin." This is true; and it is hard to see what palm of higher praise can be awarded to any philosopher.

I have always admitted myself to be under the deepest obligations to Mr. Darwin's works; and it was with the greatest reluctance, not to say repugnance, that I became one of his opponents. I have partaken of his hospitality, and have had too much experience of the charming simplicity of his manner not to be among the readiest to at once admire and envy it. It is unfortunately true that I believe Mr. Darwin to have behaved badly to me; this is too notorious to be denied; but at the same time I cannot be blind to the fact that no man can be judge in his own case, and that after all Mr. Darwin may have been right, and I wrong.

At the present moment, let me impress this latter alternative upon my mind as far as possible, and dwell only upon that side of Mr. Darwin's work and character, about which there is no difference of opinion among either his admirers or his opponents.

April 21, 1882.

PREFACE.

Contrary to the advice of my friends, who caution me to avoid all appearance of singularity, I venture upon introducing a practice, the expediency of which I will submit to the judgment of the reader. It is one which has been adopted by musicians for more than a century--to the great convenience of all who are fond of music--and I observe that within the last few years two such distinguished painters as Mr. Alma-Tadema and Mr. Hubert Herkomer have taken to it. It is a matter for regret that the practice should not have been general at an earlier date, not only among painters and musicians, but also among the people who write books. It consists in signifying the number of a piece of music, picture, or book by the abbreviation "Op." and the number whatever it may happen to be.

No work can be judged intelligently unless not only the author's relations to his surroundings, but also the relation in which the work stands to the life and other works of the author, is understood and borne in mind; nor do I know any way of conveying this information at a glance, comparable to that which I now borrow from musicians. When we see the number against a work of Beethoven, we need ask no further to be informed concerning the general character of the music. The same holds good more or less with all composers. Handel's works were not numbered--not at least his operas and oratorios. Had they been so, the significance of the numbers on Susanna and Theodora would have been at once apparent, connected as they would have been with the number on Jephthah, Handel's next and last work, in which he emphatically repudiates the influence which, perhaps in a time of self-distrust, he had allowed contemporary German music to exert over him. Many painters have dated their works, but still more have neglected doing so, and some of these have been not a little misconceived in consequence. As for authors, it is unnecessary to go farther back than Lord Beaconsfield, Thackeray, Dickens, and Scott, to feel how much obliged we should have been to any custom that should have compelled them to number their works in the order in which they were written. When we think of Shakespeare, any doubt which might remain as to the advantage of the proposed innovation is felt to disappear.

My friends, to whom I urged all the above, and more, met me by saying that the practice was doubtless a very good one in the abstract, but that no one was particularly likely to want to know in what order my books had been written. To which I answered that even a bad book which introduced so good a custom would not be without value, though the value might lie in the custom, and not in the book itself; whereon, seeing that I was obstinate, they left me, and interpreting their doing so into at any rate a modified approbation of my design, I have carried it into practice.

The edition of the 'Philosophie Zoologique' referred to in the following volume, is that edited by M. Chas. Martins, Paris, Librairie F. Savy, 24, Rue de Hautefeuille, 1873.

The edition of the 'Origin of Species' is that of

1876, unless another edition be especially named.

The italics throughout the book are generally mine, except in the quotations from Miss Seward, where they are all her own.

I am anxious also to take the present opportunity of acknowledging the obligations I am under to my friend Mr. H. F. Jones, and to other friends (who will not allow me to mention their names, lest more errors should be discovered than they or I yet know of), for the invaluable assistance they have given me while this work was going through the press. If I am able to let it go before the public with any comfort or peace of mind, I owe it entirely to the carefulness of their supervision.

I am also greatly indebted to Mr. Garnett, of the British Museum, for having called my attention to many works and passages of which otherwise I should have known nothing.

March 31, 1879.

CONTENTS.

CHAPTER XXII.

The Case of the Madeira Beetles as illustrating the Difference between the Evolution of Lamarck and of Mr. Charles Darwin--Conclusion

EVOLUTION, OLD AND NEW

CHAPTER I.

STATEMENT OF THE QUESTION. CURRENT OPINION ADVERSE TO TELEOLOGY.

Of all the questions now engaging the attention of those whose destiny has commanded them to take more or less exercise of mind, I know of none more interesting than that which deals with what is called teleology--that is to say, with design or purpose, as evidenced by the different parts of animals and plants.

The question may be briefly stated thus:--

Can we or can we not see signs in the structure of animals and plants, of something which carries with it the idea of contrivance so strongly that it is impossible for us to think of the structure, without at the same time thinking of contrivance, or design, in connection with it?

It is my object in the present work to answer this question in the affirmative, and to lead my reader to agree with me, perhaps mainly, by following the history of that opinion which is now supposed to be fatal to a purposive view of animal and vegetable organs. I refer to the theory of evolution or descent with modification.

Let me state the question more at large.

When we see organs, or living tools--for there is no well-developed organ of any living being which is not used by its possessor as an instrument or tool for the effecting of some purpose which he considers or has considered for his advantage--when we see living tools which are as admirably fitted for the work required of them, as is the carpenter's plane for planing, or the blacksmith's hammer and anvil for the hammering of iron, or the tailor's needle for sewing, what conclusion shall we adopt concerning them?

Shall we hold that they must have been designed or contrived, not perhaps by mental processes indistinguishable from those by which the carpenter's saw or the watch has been designed, but still by processes so closely resembling these that no word can be found to express the facts of the case so nearly as the word "design"? That is to say, shall we imagine that they were arrived at by a living mind as the result of scheming and contriving, and thinking (not without occasional mistakes) which of the courses open to it seemed best fitted for the occasion, or are we to regard the apparent connection between such an organ, we will say, as the eye, and the sight which is affected by it, as in no way due to the design or plan of a living intelligent being, but as caused simply by the accumulation, one upon another, of an almost infinite series of small pieces of good fortune?

In other words, shall we see something for which, as Professor Mivart has well said, "to us the word 'mind' is the least inadequate and misleading symbol," as having given to the eagle an eyesight which can pierce the sun, but which, in the night is powerless; while to the owl it has given eyes which shun even the full moon, but find a soft brilliancy in darkness? Or shall we deny that there has been any purpose or design in the fashioning of these different kinds of eyes, and see nothing to make us believe that any living being made the eagle's eye out of something which was not an eye nor anything like one, or that this living being implanted this particular eye of all others in the eagle's head, as being most in accordance with the habits of the creature, and as therefore most likely to enable it to live contentedly and leave plenitude of offspring? And shall we then go on to maintain that the eagle's eye was formed little by little by a series of accidental variations, each one of which was thrown for, as it were, with dice?

We shall most of us feel that there must have been a little cheating somewhere with these accidental variations before the eagle could have become so great a winner.

I believe I have now stated the question at issue so plainly that there can be no mistake about its nature, I will therefore proceed to show as briefly as possible what have been the positions taken in regard to it by our forefathers, by the leaders of opinion now living, and what I believe will be the next conclusion that will be adopted for any length of time by any considerable number of people.

In the times of the ancients the preponderance of opinion was in favour of teleology, though impugners were not wanting. Aristotle[1] leant towards a denial of purpose, while Plato[2] was a firm believer in design. From the days of Plato to our own times, there have been but few objectors to the teleological or purposive view of nature. If an animal had an eye, that eye was regarded as something which had been designed in order to enable its owner to see after such fashion as should be most to its advantage.

This, however, is now no longer the prevailing opinion either in this country or in Germany.

Professor Haeckel holds a high place among the leaders of German philosophy at the present day. He declares a belief in evolution and in purposiveness to be incompatible, and denies purpose in language which holds out little prospect of a compromise.

"As soon, in fact," he writes, "as we acknowledge the exclusive activity of the physico-chemical causes in living (organic) bodies as well as in so-called inanimate (inorganic) nature,"--and this is what Professor Haeckel holds we are bound to do if we accept the theory of descent with modification--"we concede exclusive dominion to that view of the universe, which we may designate as *mechanical*, and which is opposed to the teleological conception. If we compare all the ideas of the universe prevalent among different nations at different times, we can divide them all into two sharply contrasted groups--a *causal* or *mechanical*, and a *teleological* or *vitalistic*. The latter has prevailed generally in biology until now, and accordingly the animal and vegetable kingdoms have been considered as the products of a creative power, acting for a definite purpose. In the contemplation of every organism, the unavoidable conviction seemed to press itself upon us, that such a wonderful machine, so complicated an apparatus for motion as exists in the organism, could only be produced by a power analogous to, but infinitely more powerful than the power of man in the construction of his machines."[3]

A little lower down he continues:--

"*I maintain with regard to*" this "*much talked of 'purpose in nature' that it has no existence but for those persons who observe phenomena in plants and animals in the most superficial manner.* Without going more deeply into the matter, we can see at once that the rudimentary organs are a formidable obstacle to this theory. And, indeed, anyone who makes a really close study of the organization and mode of life of the various animals and plants, ... must necessarily come to the conclusion, that this 'purposiveness' no more exists than the much talked of 'beneficence' of the Creator."[4]

Professor Haeckel justly sees no alternative between, upon the one hand, the creation of independent species by a Personal God--by a "Creator," in fact, who "becomes an organism, who designs a plan, reflects upon and varies this plan, and finally forms creatures according to it, as a human architect would construct his building,"[5]--and the denial of all plan or purpose whatever. There can be no question but that he is right here.

To talk of a "designer" who has no tangible existence, no organism with which to think, no bodily mechanism with which to carry his purposes into effect; whose design is not design inasmuch as it has to contend with no impediments from ignorance or impotence, and who thus contrives but by a sort of make-believe in which there is no contrivance; who has a familiar name, but nothing beyond a name which any human sense has ever been able to perceive--this is an abuse of words--an attempt to palm off a shadow upon our understandings as though it were a substance. It is plain therefore that there must either be a designer who "becomes an organism, designs a plan, &c.," or that there can be no designer at all and hence no design.

We have seen which of these alternatives Professor Haeckel has adopted. He holds that those who accept evolution are bound to reject all "purposiveness." And here, as I have intimated, I differ from him, for reasons which will appear presently. I believe in an organic and tangible designer of every complex structure, for so long a time past, as that reasonable people will be incurious about all that occurred at any earlier time.

Professor Clifford, again, is a fair representative of opinions which are finding favour with the majority of our own thinkers. He writes:--

"There are here some words, however, which require careful definition. And first the word purpose. A thing serves a purpose when it is adapted for some end; thus a corkscrew is adapted to the end of extracting corks from bottles, and our lungs are adapted to the end of respiration. We may say that the extraction of corks is the purpose of the corkscrew, and that respiration is the purpose of the lungs, but here we shall have used the word in two different senses. A man made the corkscrew with a purpose in his mind, and he knew and intended that it should be used for pulling out corks. *But nobody made our lungs with a purpose in his mind and intended that they should be used for breathing.* The respiratory apparatus was adapted to its purpose by natural selection, namely, by the gradual preservation of better and better adaptations, and by the killing-off of the worse and imperfect adaptations."[6]

No denial of anything like design could be more explicit. For Professor Clifford is well aware that the very essence of the "Natural Selection" theory, is that the variations shall have been mainly accidental and without design of any sort, but that the adaptations of structure to need shall have come about by the accumulation, through natural selection, of any variation that *happened* to be favourable.

It will be my business on a later page not only to show that the lungs are as purposive as the corkscrew, but furthermore that if drawing corks had been a matter of as much importance to us as breathing is, the list of our organs would have been found to comprise one corkscrew at the least, and possibly two, twenty, or ten thousand; even as we see that the trowel without which the beaver cannot plaster its habitation in such fashion as alone satisfies it, is incorporate into the beaver's own body by way of a tail, the like of which is to be found in no other animal.

To take a name which carries with it a far greater authority, that of Mr. Charles Darwin. He writes:--

"It is scarcely possible to avoid comparing the eye with a telescope. We know that this instrument has been perfected by the long-continued efforts of the highest human intellects; and we naturally infer that the eye has been formed by a somewhat analogous process. But may not this inference be presumptuous? Have we any right to declare that the Creator works by intellectual powers like those of man?"[7]

Here purposiveness is not indeed denied point-blank, but the intention of the author is unmistakable, it is to refer the wonderful result to the gradual accumulation of small accidental improvements which were not due as a rule, if at all, to anything "analogous" to design.

"Variation," he says, "will cause the slight alterations;" that is to say, the slight successive variations whose accumulation results in such a marvellous structure as the eye, are caused by--variation; or in other words, they are indefinite, due to nothing that we can lay our hands upon, and therefore certainly not due to design. "Generation," continues Mr. Darwin, "will multiply them almost infinitely, and natural selection will pick out with unerring skill each improvement. Let this process go on for millions of years, and during each year on millions of individuals of many kinds; and may we not believe that a living optical instrument might be thus formed as superior to one of glass, as the works of the Creator are to those of man?"[8]

The reader will observe that the only skill--and this involves design--supposed by Mr. Darwin to be exercised in the foregoing process, is the "unerring skill" of natural selection. Natural selection, however, is, as he himself tells us, a synonym for the survival of the fittest, which last he declares to be the "more accurate" expression, and to be "sometimes" equally convenient.[9] It is clear then that he only speaks metaphorically when he here assigns "unerring skill" to the fact that the fittest individuals commonly live longest and transmit most offspring, and that he sees no evidence of design in the numerous slight successive "alterations"--or variations--which are "caused by variation."

It were easy to multiply quotations which should prove that the denial of "purposiveness" is commonly conceived to be the inevitable accompaniment of a belief in evolution. I will, however, content myself with but one more--from Isidore Geoffroy St. Hilaire.

"Whoever," says this author, "holds the doctrine of final causes, will, if he is consistent, hold also that of the immutability of species; and again, the opponent of the one doctrine will oppose the other also."[10]

Nothing can be plainer; I believe, however, that

even without quotation the reader would have recognized the accuracy of my contention that a belief in the purposiveness or design of animal and vegetable organs is commonly held to be incompatible with the belief that they have all been evolved from one, or at any rate, from not many original, and low, forms of life. Generally, however, as this incompatibility is accepted, it is not unchallenged. From time to time a voice is uplifted in protest, whose tones cannot be disregarded.

"I have always felt," says Sir William Thomson, in his address to the British Association, 1871, "that this hypothesis" (natural selection) "does not contain the true theory of evolution, if indeed evolution there has been, in biology. Sir John Herschel, in expressing a favourable judgment on the hypothesis of zoological evolution (with however some reservation in respect to the origin of man), objected to the doctrine of natural selection on the ground that it was too like the Laputan method of making books, and that it did not sufficiently take into account a continually guiding and controlling intelligence. This seems to me a most valuable and instructive criticism. *I feel profoundly convinced that the argument of design has been greatly too much lost sight of in recent zoological speculations.* Reaction against the frivolities of teleology such as are to be found in the notes of the learned commentators on Paley's 'Natural Theology,' has, I believe, had a temporary effect in turning attention from the solid and irrefragable argument so well put forward in that excellent old book. But overpoweringly strong proofs of intelligent and benevolent design lie all around us,"[11] &c. Sir William Thomson goes on to infer that all living beings depend on an ever-acting Creator and Ruler--meaning, I am afraid, a Creator who is not an organism. Here I cannot follow him, but while gladly accepting his testimony to the omnipresence of intelligent design in almost every structure, whether of animal or plant, I shall content myself with observing the manner in which plants and animals act and with the consequences that are legitimately deducible from their action.

FOOTNOTES:

[1] See note to Mr. Darwin, Historical Sketch, &c., 'Origin of Species, p. xiii. ed. 1876, and Arist. 'Physicæ Auscultationes,' lib. ii. cap. viii. s. 2.

[2] See Phædo and Timæus.

[3] 'History of Creation,' vol. i. p. 18 (H. S. King and Co., 1876).

[4] Ibid. p. 19.

[5] 'History of Creation,' vol. i. p. 73 (H. S. King and Co., 1876).

[6] 'Fortnightly Review,' new series, vol. xviii. p. 795.

[7] 'Origin of Species,' p. 146, ed. 1876.

[8] 'Origin of Species,' p. 146, ed. 1876.

[9] Page 49.

[10] 'Vie et Doctrine scientifique d'Étienne Geoffroy St. Hilaire,' by Isidore Geoffroy St. Hilaire. Paris, 1847, p. 344.

[11] Address to the British Association, 1871.

CHAPTER II

THE TELEOLOGY OF PALEY AND THE THEOLOGIANS.

Let us turn for a while to Paley, to whom Sir W. Thomson has referred us. His work should be so well known that an apology is almost due for quoting it, yet I think it likely that at least nine out of ten of my readers will (like myself till reminded of it by Sir W. Thomson's address) have forgotten its existence.

"In crossing a heath," says Paley, "suppose I pitched my foot against a stone, and were asked how the stone came to be there; I might possibly answer that for anything I knew to the contrary, it had lain there for ever; nor would it perhaps be very easy to show the absurdity of this answer. But suppose I had found a *watch* upon the ground, and it should be inquired how the watch happened to be in that place; I should hardly think of the answer I had before given--that for anything I knew the watch might have been always there. Yet, why should not this answer serve for the watch as well as for the stone? Why is it not as admissible in the second case as in the first? For this reason, and for no other, viz. that when we come to inspect the watch, we perceive (what we could not discover in the stone) that its several parts are framed and put together for a purpose, e. g. that they are so formed and adjusted as to produce motion, and that motion so regulated as to point out the hour of the day: that if the different parts had been differently shaped from what they are, of a different size from what they are, or placed after any other manner, or in any other order, than that in which they are placed, either no motion at all would have been carried on in the machine, or none that would have answered the use which is now served by it. To reckon up a few of the plainest of these parts, and of their offices all tending to one result: we see a cylindrical box containing a coiled elastic spring, which, by its endeavours to relax itself, turns round the box. We next observe a flexible chain (artificially wrought for the sake of flexure) communicating the action of the spring from the box to the fusee. We then find a series of wheels the teeth of which catch in, and apply to each other, conducting the motion from the fusee to the balance, and from the balance to the pointer; and at the same time by the size and shape of those wheels so regulating the motion as to terminate in causing an index, by an equable and measured progression, to pass over a given space in a given time. We take notice that the wheels are made of brass in order to keep them from rust; the springs of steel, no other metal being so elastic; that over the face of the watch there is placed a glass, a material employed on no other part of the work, but in the room of which if there had been any other than a transparent substance, the hour could not have been observed without opening the case. This mechanism being observed, ... the inference, we think, is inevitable that the watch must have had a maker; that there must have existed, at *some time, and at some place or other, an artificer* or artificers who formed it for the purpose which we find it actually to answer; who comprehended its construction and designed its use."[12]

"That an animal is a machine, is a proposition neither correctly true nor wholly false.... I contend that there is a mechanism in animals; that this mechanism is as properly such, as it is in machines made by art; that this mechanism is intelligible and certain; that it is not the less so because it often begins and terminates with something which is not mechanical; that wherever it is intelligible and certain, it demonstrates intention and contrivance, as well in the works of nature as in those of art; and that it is the best demonstration which either can afford."[13]

There is only one legitimate inference deducible from these premises if they are admitted as sound, namely, that there must have existed "*at some time, and in some place, an artificer*" who formed the animal mechanism after much the same mental processes of observation, endeavour, successful contrivance, and after a not wholly unlike succession of bodily actions, as those with which a watchmaker has made a watch. Otherwise the conclusion is impotent, and the whole argument becomes a mere juggle of words.

"Now, supposing or admitting," continues Paley, "that we know nothing of the proper internal constitution of a gland, or of the mode of its acting upon the blood; then our situation is precisely like that of an unmechanical looker-on who stands by a stocking loom, a corn mill, a carding machine, or a threshing machine, at work, the fabric and mechanism of which, as well as all that passes within, is hidden from his sight by the outside case; or if seen, would be too complicated for his uninformed, uninstructed understanding to comprehend. And what is that situation? This spectator, ignorant as he is, sees at one end a material enter the machine, as unground grain the mill, raw cotton the carding machine, sheaves of unthreshed corn the threshing machine, and when he casts his eye to the other end of the apparatus, he sees the material issuing from it in a new state and what is more, a state manifestly adapted for its future uses: the grain in meal fit for the making of bread, the wool in rovings fit for the spinning into threads, the sheaf in corn fit for the mill. Is it necessary that this man, in order to be convinced that design, that intention, that contrivance has been employed about the machine, should be allowed to pull it to pieces, should be enabled to examine the parts separately, explore their action upon one another, or their operation, whether simultaneous or successive, upon the material which is presented to them? He may long to do this to satisfy his curiosity; he may desire to do it to improve his theoretic knowledge; ... but for the purpose of ascertaining the existence of counsel and design in the formation of the machine, he wants no such intromission or privity. The effect upon the material, the change produced in it, the utility of the change for future applications, abundantly testify, be the concealed part of the machine, or of its construction, what it will, *the hand and agency of a contriver.*"[14]

This is admirably put, but it will apply to the mechanism of animal and vegetable bodies only, if it is used to show that they too must have had a contriver who has a hand, or something tantamount to one; who does act; who, being a contriver, has what all other contrivers must have, if they are to be called contrivers--a body which can suffer more or less pain or chagrin if the contrivance is unsuccessful. If this is what Paley means, his argument is indeed irrefragable; but if he does not intend this, his words are frivolous, as so clear and acute a reasoner must have perfectly well known.

Whether Paley's argument will prove a source of lasting strength to himself or no, is a point which my readers will decide presently; but I am very clear about its usefulness to my own position. I know few writers whom I would willingly quote more largely, or from whom I find it harder to leave off quoting when I have once begun. A few more passages, however, must suffice.

"I challenge any man to produce in the joints and pivots of the most complicated or the most flexible machine that ever was contrived, a construction *more artificial*" (here we have it again), "or more evidently artificial than the human neck. Two things were to be done. The head was to have the power of bending forward and backward as in the act of nodding, stooping, looking upwards or downwards; and at the same time of turning itself round upon the body to a certain extent, the quadrant, we will say, or rather perhaps a hundred and twenty degrees of a circle. For these two purposes two distinct contrivances are employed. First the head rests immediately upon the uppermost part of the vertebra, and is united to it by a hinge-joint; upon this joint the head plays freely backward and forward as far either way as is necessary or as the ligaments allow, which was the first thing required.

"But then the rotatory motion is thus unprovided for; therefore, secondly, to make the head capable of this a further mechanism is introduced, not between the head and the uppermost bone of the neck, where the hinge is, but between that bone and the next underneath it. It is a mechanism resembling a tenon and mortise. This second or uppermost bone but one has what the anatomists call a process, viz. a projection somewhat similar in size and shape to a tooth, which tooth, entering a corresponding hollow socket in the bone above it, forms a pivot or axle, upon which that upper bone, together with the head which it supports, turns freely in a circle, and as far in the circle as the attached muscles permit the head to turn. Thus are both motions perfect without interfering with each other. When we nod the head we use the hinge-joint, which lies between the head and the first bone of the neck. When we turn the head round, we use the tenon and mortise, which runs between the first bone of the neck and the second. We see the same contrivance and the same principle employed in the frame or mounting of a telescope. It is occasionally requisite that the object end of the instrument be moved up and down as well as horizontally or equatorially. For the vertical motion there is a hinge upon which the telescope plays, for the horizontal or equatorial motion, an axis upon which the telescope and the hinge turn round together. And this is exactly the mechanism which is applied to the action of the head, nor will anyone here doubt of the existence of counsel and design, except it be by that debility of mind which can trust to its own reasonings in nothing."[15]

"The patella, or knee-pan, is a curious little bone; in its form and office unlike any other bone in the body. It is circular, the size of a crown-piece, pretty thick, a little convex on both sides, and covered with a smooth cartilage. It lies upon the front of the knee, and the powerful tendons by which the leg is brought forward pass through it (or rather make it a part of their continuation) from their origin in the thigh to their insertion in the tibia. It protects both the tendon and the joint from any injury which either might suffer by the rubbing of one against the other, or by the pressure of unequal surfaces. It also gives to the tendons a very considerable mechanical advantage by altering the line of their direction, and by advancing it farther out of the centre of motion; and this upon the principles of the resolution of force, upon which all machinery is founded. These are its uses. But what is most observable in it is that it appears to be supplemental, as it were, to the frame; added, as it should almost seem, afterwards; not quite necessary, but very convenient. It is separate from the other bones; that is, it is not connected with any other bones by the common mode of union. It is soft, or hardly formed in infancy; and is produced by an ossification, of the inception or progress of which no account can be given from the structure or exercise of the part."[16]

It is positively painful to me to pass over Paley's description of the joints, but I must content myself with a single passage from this admirable chapter.

"The joints, or rather the ends of the bones which form them, display also in their configuration another use. The nerves, blood-vessels, and tendons which are necessary to the life, or for the motion of the limbs, must, it is evident in their way from the trunk of the body to the place of their destination, travel over the moveable joints; and it is no less evident that in this part of their course they will have from sudden motions, and from abrupt changes of curvature, to encounter the danger of compression, attrition, or laceration. To guard fibres so tender against consequences so injurious, their path is in those parts protected with peculiar care; and that by a provision in the figure of the bones themselves. The nerves which supply the fore arm, especially the inferior cubital nerves, are at the elbow conducted by a kind of covered way, between the condyle, or rather under the inner extuberances, of the bone which composes the upper part of the arm. At the knee the extremity of the thigh-bone is divided by a sinus or cliff into two heads or protuberances; and these heads on the back part stand out beyond the cylinder of the bone. Through the hollow which lies between the hind parts of these two heads, that is to say, under the ham, between the ham strings, and within the concave recess of the bone formed by the extuberances on either side; in a word, along a defile between rocks pass the great vessels and nerves which go to the leg. Who led these vessels by a road so defended and secured? In the joint at the shoulder, in the edge of the cup which receives the head of the bone, is a notch which is covered at the top with a ligament. Through this hole thus guarded the blood-vessels steal to their destination in the arm instead of mounting over the edge of the concavity."[17]

"What contrivance can be more mechanical than the following, viz.: a slit in one tendon to let another tendon pass through it? This structure is found in the tendons which move the toes and fingers. The long tendon, as it is called in the foot, which bends the first joint of the toe, passes through the short tendon which bends the second joint; which course allows to the sinews more liberty and a more commodious action than it would otherwise have been capable of exerting. There is nothing, I believe, in a silk or cotton mill, in the belts or straps or ropes by which the motion is communicated from one part of the machine to another that is more artificial, or more evidently so, than this perforation.

"The next circumstance which I shall mention under this head of muscular arrangement, is so decidedly a mark of intention, that it always appeared to me to supersede in some measure the necessity of seeking for any other observation upon the subject; and that circumstance is the tendons which pass from the leg to the foot being bound down by a ligament at the ankle, the foot is placed at a considerable angle with the leg. It is manifest, therefore, that flexible strings passing along the interior of the angle, if left to themselves, would, when stretched, start from it. The obvious" (and it must not be forgotten that the preventive *was* obvious) "preventive is to tie them down. And this is done in fact. Across the instep, or rather just above it, the anatomist finds a strong ligament, under which the tendons pass to the foot. The effect of the ligament as a bandage can be made evident to the senses, for if it be cut the tendons start up. The simplicity, yet the clearness of this contrivance, its exact resemblance to established resources of art, place it amongst the most indubitable manifestations of design with which we are acquainted."

Then follows a passage which is interesting, as being the earliest attempt I know of to bring forward an argument against evolution, which was, even in Paley's day, called "Darwinism," after Dr. Erasmus Darwin its propounder.[18] The argument, I mean, which is drawn from the difficulty of accounting for the incipiency of complex structures. This has been used with greater force by the Rev. J. J. Murphy, Professor Mivart, and others, against that (as I believe) erroneous view of evolution which is now generally received as Darwinism.

"There is also a further use," says Paley, "to be made of this present example, and that is as it precisely contradicts the opinion, that the parts of animals may have been all formed by what is called appetency, i. e. endeavour, perpetuated and imperceptibly working its effect through an incalculable series of generations. We have here no endeavour, but the reverse of it; a constant resistency and reluctance. The endeavour is all the other way. The pressure of the ligament constrains the tendons; the tendons react upon the pressure of the ligament. It is impossible that the ligament should ever have been generated by the exercise of the tendons, or in the course of that exercise, forasmuch as the force of the tendon perpendicularly resists the fibre which confines it, and is constantly endeavouring not to form but to rupture and displace the threads of which the ligament is composed."[19]

This must suffice.

"True theories," says M. Flourens, inspired by a passage from Fontenelle, which he proceeds to quote, "true theories make themselves," they are not made, but are born and grow; they cannot be stopped from insisting upon their vitality by anything short of intellectual violence, nor will a little violence only suffice to kill them. "True theories," he continues, "are but the spontaneous mental coming together of facts, which have combined with one another by virtue only of their own natural affinity."[20]

When a number of isolated facts, says Fontenelle, take form, group themselves together coherently, and present the mind so vividly with an idea of their interdependence and mutual bearing upon each other, that no matter how violently we tear them asunder they insist on coming together again; then, and not till then, have we a theory.

Now I submit that there is hardly one of my readers who can be considered as free from bias or prejudice, who will not feel that the idea of design--or perception by an intelligent living being, of ends to be obtained and of the means of obtaining them--and the idea of the tendons of the foot and of the ligament which binds them down, come together so forcibly, that no matter how strongly Professors Haeckel and Clifford and Mr. Darwin may try to separate them, they are no sooner pulled asunder than they straightway fly together again of themselves.

I shall argue, therefore, no further upon this head, but shall assume it as settled, and shall proceed at once to the consideration that next suggests itself.

FOOTNOTES:

[12] 'Natural Theology,' ch. i. § 1.

[13] Ch. vii.

[14] Ch. vii.

[15] 'Natural Theology.' ch. viii.

[16] 'Natural Theology,' ch. viii.

[17] 'Natural Theology,' ch. viii.

[18] "What!" says Coleridge, in a note on Stillingfleet, to which Mr. Garnett, of the British Museum, has kindly called my attention, "Did Sir

Walter Raleigh believe that a male and female ounce (and if so why not two tigers and lions, &c.?) would have produced in course of generations a cat, or a cat a lion? This is Darwinising with a vengeance."--See 'Athenæum,' March 27, 1875, p. 423.

[19] 'Natural Theology,' ch. ix.

[20] "La vraie théorie n'est que l'enchaînement naturel des faits, qui dès qu'ils sont assez nombreux, se touchent, et se lient, les uns aux autres par leur seule vertu propre."--Flourens, 'Buffon, Hist. de ses Travaux.' Paris, 1844, p. 82.

CHAPTER III.

IMPOTENCE OF PALEY'S CONCLUSION. THE TELEOLOGY OF THE EVOLUTIONIST.

Though the ideas of design, and of the foot, have come together in our minds with sufficient spontaneity, we yet feel that there is a difference--and a wide difference if we could only lay our hands upon it--between the design and manufacture of the ligament and tendons of the foot on the one hand, and on the other the design, manufacture, and combination of artificial strings, pieces of wood, and bandages, whereby a model of the foot might be constructed.

If we conceive of ourselves as looking

simultaneously upon a real foot, and upon an admirably constructed artificial one, placed by the side of it, the idea of design, and design by an intelligent living being with a body and soul (without which, as has been already insisted on, the use of the word design is delusive), will present itself strongly to our minds in connection both with the true foot, and with the model; but we find another idea asserting itself with even greater strength, namely, that the design of the true foot is far more intricate, and yet is carried into execution in far more masterly manner than that of the model. We not only feel that there is a wider difference between the ability, time, and care which have been lavished on the real foot and upon the model, than there is between the skill and the time taken to produce Westminster Abbey, and that bestowed upon a gingerbread cake stuck with sugar plums so as to represent it, but also that these two objects must have been manufactured on different principles. We do not for a moment doubt that the real foot was designed, but we are so astonished at the dexterity of the designer that we are at a loss for some time to think who could have designed it, where he can live, in what manner he studied, for how long, and by what processes he carried out his design, when matured, into actual practice. Until recently it was thought that there was no answer to many of these questions, more especially to those which bear upon the mode of manufacture. For the last hundred years, however, the importance of a study has been recognized which does actually reveal to us in no small degree the processes by which the human foot is manufactured, so that in the endeavour to lay our hands upon the points of difference between the kind of design with which the foot itself is designed, and the design of the model, we turn naturally to the guidance of those who have made this study their specialty; and a very wide difference does this study, embryology, at once reveal to us.

Writing of the successive changes through which each embryo is forced to pass, the late Mr. G. H. Lewes says that "none of these phases have any adaptation to the future state of the animal, but are in positive contradiction to it or are simply purposeless; whereas all show stamped on them the unmistakable characters of *ancestral* adaptation, and the progressions of organic evolution. What does the fact imply? There is not a single known example of a complex organism which is not developed out of simpler forms. Before it can attain the complex structure which distinguishes it, there must be an evolution of forms similar to those which distinguish the structure of organisms lower in the series. On the hypothesis of a plan which prearranged the organic world, nothing could be more unworthy of a supreme intelligence than this inability to construct an organism at once, without making several previous tentative efforts, undoing to-day what was so carefully done yesterday, and *repeating for centuries the same tentatives in the same succession*. Do not let us blink this consideration. There is a traditional phrase much in vogue among the anthropomorphists, which arose naturally enough from a tendency to take human methods as an explanation of the Divine--a phrase which becomes a sort of argument--'The Great Architect.' But if we are to admit the human point of view, a glance at the facts of embryology must produce very uncomfortable reflections. For what should we say to an architect who was unable, or being able was obstinately unwilling, to erect a palace except by first using his materials in the shape of a hut, then pulling them down and rebuilding them as a cottage, then adding story to story and room to room, *not* with any reference to the ultimate purposes of the palace, but wholly with reference to the way in which houses were constructed in ancient times? What should we say to the architect who could not form a museum out of bricks and mortar, but was forced to begin as if going to construct a mansion, and after proceeding some way in this direction, altered his plan into a palace, and that again into a museum? Yet this is the sort of succession on which organisms are constructed. The fact has long been familiar; how has it been reconciled with infinite wisdom? Let the following passage answer for a thousand:--'The embryo is nothing like the miniature of the adult. For a long while the body in its entirety and in its details, presents the strangest of spectacles. Day by day and hour by hour, the aspect of the scene changes, and this instability is exhibited by the most essential parts no less than by the accessory parts. One would say that nature feels her way, and only reaches the goal after many times missing the path' (on dirait que la nature tâtonne et ne conduit son oeuvre à bon fin, qu'après s'être souvent trompée)."[21]

The above passage does not, I think, affect the evidence for design which we adduced in the preceding chapter. However strange the process of manufacture may appear, when the work comes to be turned out the design is too manifest to be doubted.

If the reader were to come upon some lawyer's deed which dealt with matters of such unspeakable intricacy, that it baffled his imagination to conceive how it could ever have been drafted, and if in spite of this he were to find the intricacy of the provisions to be made, exceeded only by the ease and simplicity with which the deed providing for them was found to work in practice; and after this, if he were to discover that the deed, by whomsoever drawn, had nevertheless been drafted upon principles which at first seemed very foreign to any according to which he was in the habit of drafting deeds himself, as for example, that the draftsman had begun to draft a will as a marriage settlement, and so forth--yet an observer would not, I take it, do either of two things. He would not in the face of the result deny the design, making himself judge rather of the method of procedure than of the achievement. Nor yet after insisting in the manner of Paley, on the wonderful proofs of intention and on the exquisite provisions which were to be found in every syllable--thus leading us up to the highest pitch of expectation--would he present us with such an impotent conclusion as that the designer, though a living person and a true designer, was yet immaterial and intangible, a something, in fact, which proves to be a nothing: an omniscient and omnipotent vacuum.

Our observer would feel he need not have been at such pains to establish his design if this was to be the upshot of his reasoning. He would therefore admit the design, and by consequence the designer, but would probably ask a little time for reflection before he ventured to say who, or what, or where the designer was. Then gaining some insight into the manner in which the deed had been drawn, he would conclude that the draftsman was a specialist who had had long practice in this particular kind of work, but who now worked almost as it might be said automatically and without consciousness, and found it difficult to depart from a habitual method of procedure.

We turn, then, on Paley, and say to him: "We have admitted your design and your designer. Where is he? Show him to us. If you cannot show him to us as flesh and blood, show him as flesh and sap; show him as a living cell; show him as protoplasm. Lower than this we should not fairly go; it is not in the bond or *nexus* of our ideas that something utterly inanimate and inorganic should scheme, design, contrive, and elaborate structures which can make mistakes: it may elaborate low unerring things, like crystals, but it cannot elaborate those which have the power to err. Nevertheless, we will commit such abuse with our understandings as to waive this point, and we will ask you to show him to us as air which, if it cannot be seen, yet can be felt, weighed, handled, transferred from place to place, be judged by its effects, and so forth; or if this may not be, give us half a grain of hydrogen, diffused through all space and invested with some of the minor attributes of matter; or if you cannot do this, give us an imponderable like electricity, or even the higher mathematics, but give us something or throw off the mask and tell us fairly out that it is your paid profession to hoodwink us on this matter if you can, and that you are but doing your best to earn an honest living."

We may fancy Paley as turning the tables upon us and as saying: "But you too have admitted a designer--you too then must mean a designer with a body and soul, who must be somewhere to be found in space, and who must live in time. Where is this your designer? Can you show him more than I can? Can you lay your finger on him and demonstrate him so that a child shall see him and know him, and find what was heretofore an isolated idea concerning him, combine itself instantaneously with the idea of the designer, we will say, of the human foot, so that no power on earth shall henceforth tear those two ideas asunder? Surely if you cannot do this, you too are trifling with words, and abusing your own mind and that of your reader. Where, then, is your designer of man? Who made him? And where, again, is your designer of beasts and birds, of fishes, and of plants?"

Our answer is simple enough; it is that we can and do point to a living tangible person with flesh, blood, eyes, nose, ears, organs, senses, dimensions, who did of his own cunning after infinite proof of every kind of hazard and experiment scheme out, and fashion each organ of the human body. This is the person whom we claim as the designer and artificer of that body, and he is the one of all others the best fitted for the task by his antecedents, and his practical knowledge of the requirements of the case--for he is man himself.

Not man, the individual of any given generation, but man in the entirety of his existence from the dawn of life onwards to the present moment. In like manner we say that the designer of all organisms is so incorporate with the organisms themselves--so lives, moves, and has its being in those organisms, and is so one with them--they in it, and it in them--that it is more consistent with reason and the common use of words to see the designer of each living form in the living form itself, than to look for its designer in some other place or person.

Thus we have a third alternative presented to us.

Mr. Charles Darwin and his followers deny design, as having any appreciable share in the formation of organism at all.

Paley and the theologians insist on design, but upon a designer outside the universe and the organism.

The third opinion is that suggested in the first instance, and carried out to a very high degree of development by Buffon. It was improved, and, indeed, made almost perfect by Dr. Erasmus Darwin, but too much neglected by him after he had put it forward. It was borrowed, as I think we may say with some confidence, from Dr. Darwin by Lamarck, and was followed up by him ardently thenceforth, during the remainder of his life, though somewhat less perfectly comprehended by him than it had been by Dr. Darwin. It is that the design which has designed organisms, has resided within, and been embodied in, the organisms themselves.

With but a very little change in the present signification of words, the question resolves itself into this.

Shall we see God henceforth as embodied in all living forms; as dwelling in them; as being that power in them whereby they have learnt to fashion themselves, each one according to its ideas of its own convenience, and to make itself not only a microcosm, or little world, but a little unwritten history of the universe from its own point of view into the bargain? From everlasting, in time past, only in so far as life has lasted; invisible, only in so far as the ultimate connection between the will to do and the thing which does is invisible; imperishable, only in so far as life as a whole is imperishable; omniscient and omnipotent, within the limits only of a very long and large experience, but ignorant and impotent in respect of all else--limited in all the above respects, yet even so incalculably vaster than anything that we can conceive?

Or shall we see God as we were taught to say we saw him when we were children--as an artificial and violent attempt to combine ideas which fly asunder and asunder, no matter how often we try to force them into combination?

"The true mainspring of our existence," says Buffon, "lies not in those muscles, veins, arteries, and nerves, which have been described with so much minuteness, it is to be found in the more hidden forces which are not bounden by the gross mechanical laws which we would fain set over them. Instead of trying to know these forces by their effects, we have endeavoured to uproot even their very idea, so as to banish them utterly from philosophy. But they return to us and with renewed vigour; they return to us in gravitation, in chemical affinity, in the phenomena of electricity, &c. Their existence rests upon the clearest evidence; the omnipresence of their action is indisputable, but that action is hidden away from our eyes, and is a matter of inference only; we cannot actually see them, therefore we find difficulty in admitting that they exist; we wish to judge of everything by its exterior; we imagine that the exterior is the whole, and deeming that it is not permitted us to go beyond it, we neglect all that may enable us to do so."[22]

Or may we not say that the unseen parts of God are those deep buried histories, the antiquity and the repeatedness of which go as far beyond that of any habit handed down to us from our earliest protoplasmic ancestor, as the distance of the remotest star in space transcends our distance from the sun?

By vivisection and painful introspection we can rediscover many a long buried history--rekindling that sense of novelty in respect of its action, whereby we can alone become aware of it. But there are other remoter histories, and more repeated thoughts and actions, before which we feel so powerless to reawaken fresh interest concerning them, that we give up the attempt in despair, and bow our heads, overpowered by the sense of their immensity. Thus our inability to comprehend God is coextensive with our difficulty in going back upon the past--and our sense of him is a dim perception of our own vast and now inconceivably remote history.

FOOTNOTES:

[21] Quatrefages, 'Metamorphoses de l'Homme et des Animaux,' 1862, p. 42; G. H. Lewes, 'Physical Basis of Mind,' 1877, p. 83.

[22] Tom. ii. p. 486, 1794.

CHAPTER IV.

FAILURE OF THE FIRST EVOLUTIONISTS TO SEE THEIR POSITION AS TELEOLOGICAL.

It follows necessarily from the doctrine of Dr. Erasmus Darwin and Lamarck, if not from that of Buffon himself, that the greater number of organs are as purposive to the evolutionist as to the theologian, and far more intelligibly so. Circumstances, however, prevented these writers from acknowledging this fact to the world, and perhaps even to themselves. Their *crux* was, as it still is to so many evolutionists, the presence of rudimentary organs, and the processes of embryological development. They would not admit that rudimentary and therefore useless organs were designed by a Creator to take their place once and for ever as part of a scheme whose main idea was, that every animal structure was to serve some useful end in connection with its possessor.

This was the doctrine of final causes as then commonly held; in the face of rudimentary organs it was absurd. Buffon was above all things else a plain matter of fact thinker, who refused to go far beyond the obvious. Like all other profound writers, he was, if I may say so, profoundly superficial. He felt that the aim of research does not consist in the knowing this or that, but in the easing of the desire to know or understand more completely--in the peace of mind which passeth all understanding. His was the perfection of a healthy mental organism by which over effort is felt instinctively to be as vicious and contemptible as indolence. He knew this too well to know the grounds of his knowledge, but we smaller people who know it less completely, can see that such felicitous instinctive tempering together of the two great contradictory principles, love of effort and love of ease, has underlain every step of all healthy growth through all conceivable time. Nothing is worth looking at which is seen either too obviously or with too much difficulty. Nothing is worth doing or well done which is not done fairly easily, and some little deficiency of effort is more pardonable than any very perceptible excess; for virtue has ever erred rather on the side of self-indulgence than of asceticism, and well-being has ever advanced through the pleasures rather than through austerity.

According to Buffon, then--as also according to Dr. Darwin, who was just such another practical and genial thinker, and who was distinctly a pupil of Buffon, though a most intelligent and original one--

if an organ after a reasonable amount of inspection appeared to be useless, it was to be called useless without more ado, and theories were to be ordered out of court if they were troublesome. In like manner, if animals bred freely *inter se* before our eyes, as for example the horse and ass, the fact was to be noted, but no animals were to be classed as capable of interbreeding until they had asserted their right to such classification by breeding with tolerable certainty. If, again, an animal looked as if it felt, that is to say, if it moved about pretty quickly or made a noise, it must be held to feel; if it did neither of these things, it did not look as if it felt and therefore it must be said not to feel. *De non apparentibus et non existentibus eadem est lex* was one of the chief axioms of their philosophy; no writers have had a greater horror of mystery or of ideas that have not become so mastered as to be, or to have been, superficial. Lamarck was one of those men of whom I believe it has been said that they have brain upon the brain. He had his theory that an animal could not feel unless it had a nervous system, and at least a spinal marrow--and that it could not think at all without a brain--all his facts, therefore, have to be made to square with this. With Buffon and Dr. Darwin we feel safe that however wrong they may sometimes be, their conclusions have always been arrived at on that fairly superficial view of things in which, as I have elsewhere said, our nature alone permits us to be comforted.

To these writers, then, the doctrine of final causes for rudimentary organs was a piece of mystification and an absurdity; no less fatal to any such doctrine were the processes of embryological development. It was plain that the commonly received teleology must be given up; but the idea of design or purpose was so associated in their minds with theological design that they avoided it altogether. They seem to have forgotten that an internal teleology is as much teleology as an external one; hence, unfortunately, though their whole theory of development is intensely purposive, it is the fact rather than the name of teleology which has hitherto been insisted upon, even by the greatest writers on evolution--the name having been denied even by those who were most insisting on the thing itself.

It is easy to understand the difficulty felt by the fathers of evolution when we remember how much had to be seen before the facts could lie well before them. It was necessary to attain, firstly, to a perception of the unity of person between parents and offspring in successive generations; secondly, it must be seen that an organism's memory goes back for generations beyond its birth, to the first beginnings in fact, of which we know anything whatever; thirdly, the latency of that memory, as of memory generally till the associated ideas are reproduced, must be brought to bear upon the facts of heredity; and lastly, the unconsciousness with which habitual actions come to be performed, must be assigned as the explanation of the unconsciousness with which we grow and discharge most of our natural functions.

Buffon was too busy with the fact that animals descended with modification at all, to go beyond the development and illustration of this great truth. I doubt whether he ever saw more than the first, and that dimly, of the four considerations above stated.

Dr. Darwin was the first to point out the first two considerations with some clearness, but he can hardly be said to have understood their full importance: the two latter ideas do not appear to have occurred to him.

Lamarck had little if any perception of any one of the four. When, however, they are firmly seized and brought into their due bearings one upon another, the facts of heredity become as simple as those of a man making a tobacco pipe, and rudimentary organs are seen to be essentially of the same character as the little rudimentary protuberance at the bottom of the pipe to which I referred in 'Erewhon.'[23]

These organs are now no longer useful, but they once were so, and were therefore once purposive, though not so now. They are the expressions of a bygone usefulness; sayings, as it were, about which there was at one time infinite wrangling, as to what both the meaning and the expression should best be, so that they then had living significance in the mouths of those who used them, though they have become such mere shibboleths and cant formulæ to ourselves that we think no more of their meaning than we do of Julius Cæsar in the month of July. They continue to be reproduced through the force of habit, and through indisposition to get out of any familiar groove of action until it becomes too unpleasant for us to remain in it any longer. It has long been felt that embryology and rudimentary structures indicated community of descent. Dr.

Darwin and Lamarck insisted on this, as have all subsequent writers on evolution; but the explanation of why and how the structures come to be repeated--namely, that they are simply examples of the force of habit--can only be perceived intelligently by those who admit so much unity between parents and offspring that the self-development of the latter can be properly called habitual (as being a repetition of an act by one and the same individual), and can only be fully sympathized with by those who recognize that if habit be admitted as the key to the fact at all, the unconscious manner in which the habit comes to be repeated is only of a piece with all our other observations concerning habit. For the fuller development of the foregoing, I must refer the reader to my work 'Life and Habit.'

The purposiveness, which even Dr. Darwin, and Lamarck still less, seem never to have quite recognized in spite of their having insisted so much on what amounts to the same thing, now comes into full view. It is seen that the organs external to the body, and those internal to it are, the second as much as the first, things which we have made for our own convenience, and with a prevision that we shall have need of them; the main difference between the manufacture of these two classes of organs being, that we have made the one kind so often that we can no longer follow the processes whereby we make them, while the others are new things which we must make introspectively or not at all, and which are not yet so incorporate with our vitality as that we should think they grow instead of being manufactured. The manufacture of the tool, and the manufacture of the living organ prove therefore to be but two species of the same genus, which, though widely differentiated, have descended as it were from one common filament of desire and inventive faculty. The greater or less complexity of the organs goes for very little. It is only a question of the amount of intelligence and voluntary self-adaptation which we must admit, and this must be settled rather by an appeal to what we find in organism, and observe concerning it, than by what we may have imagined *à priori.*

Given a small speck of jelly with some kind of circumstance-suiting power, some power of slightly varying its actions in accordance with slightly varying circumstances and desires--given such a jelly-speck with a power of assimilating other matter, and thus, of reproducing itself, given also that it should be possessed of a memory, and we can show how the whole animal world can have descended it may be from an amoeba without interference from without, and how every organ in every creature is designed at first roughly and tentatively but finally fashioned with the most consummate perfection, by the creature which has had need of that organ, which best knew what it wanted, and was never satisfied till it had got that which was the best suited to its varying circumstances in their entirety. We can even show how, if it becomes worth the Ethiopian's while to try and change his skin, or the leopard's to change his spots, they can assuredly change them within a not unreasonable time and adapt their covering to their own will and convenience, and to that of none other; thus what is commonly conceived of as direct creation by God is moved back to a time and space inconceivable in their remoteness, while the aim and design so obvious in nature are shown to be still at work around us, growing ever busier and busier, and advancing from day to day both in knowledge and power.

It was reserved for Mr. Darwin and for those who have too rashly followed him to deny purpose as having had any share in the development of animal and vegetable organs; to see no evidence of design in those wonderful provisions which have been the marvel and delight of observers in all ages. The one who has drawn our attention more than perhaps any other living writer to those very marvels of coadaptation, is the foremost to maintain that they are the result not of desire and design, either within the creature or without it, but of blind chance, working no whither, and due but to the accumulation of innumerable lucky accidents.

"There are men," writes Professor Tyndall in the 'Nineteenth Century,' for last November, "and by no means the minority, who, however wealthy in regard to facts, can never rise into the region of principles; and they are sometimes intolerant of those that can. They are formed to plod meritoriously on in the lower levels of thought; unpossessed of the pinions necessary to reach the heights, they cannot realize the mental act--the act of inspiration it might well be called--by which a man of genius, after long pondering and proving, reaches a theoretic conception which unravels and illuminates the tangle of centuries of observation and experiment. There are minds, it may be said in passing, who, at the present moment, stand in this relation to Mr. Darwin."

The more rhapsodical parts of the above must go for what they are worth, but I should be sorry to think that what remains conveyed a censure which might fall justly on myself. As I read the earlier part of the passage I confess that I imagined the conclusion was going to be very different from what it proved to be. Fresh from the study of the older men and also of Mr. Darwin himself, I failed to see that Mr. Darwin had "unravelled and illuminated" a tangled skein, but believed him, on the contrary, to have tangled and obscured what his predecessors had made in great part, if not wholly, plain. With the older writers, I had felt as though in the hands of men who wished to understand themselves and to make their reader understand them with the smallest possible exertion. The older men, if not in full daylight, at any rate saw in what quarter of the sky the dawn was breaking, and were looking steadily towards it. It is not they who have put their hands over their own eyes and ours, and who are crying out that there is no light, but chance and blindness everywhere.

FOOTNOTES:

[23] Page 210, first edition.

CHAPTER V.

THE TELEOLOGICAL EVOLUTION OF ORGANISM--THE PHILOSOPHY OF THE UNCONSCIOUS.

I have stated the foregoing in what I take to be an extreme logical development, in order that the reader may more easily perceive the consequences of those premises which I am endeavouring to re-establish. But it must not be supposed that an animal or plant has ever conceived the idea of some organ widely different from any it was yet possessed of, and has set itself to design it in detail and grow towards it.

The small jelly-speck, which we call the amoeba, has no organs save what it can extemporize as occasion arises. If it wants to get at anything, it thrusts out part of its jelly, which thus serves it as an arm or hand: when the arm has served its purpose, it is absorbed into the rest of the jelly, and has now to do the duty of a stomach by helping to wrap up what it has just purveyed. The small round jelly-speck spreads itself out and envelops its food, so that the whole creature is now a stomach, and nothing but a stomach. Having digested its food, it again becomes a jelly-speck, and is again ready to turn part of itself into hand or foot as its next convenience may dictate. It is not to be believed that such a creature as this, which is probably just sensitive to light and nothing more, should be able to form a conception of an eye and set itself to work to grow one, any more than it is believable that he who first observed the magnifying power of a dew drop, or even he who first constructed a rude lens, should have had any idea in his mind of Lord Rosse's telescope with all its parts and appliances. Nothing could be well conceived more foreign to experience and common sense. Animals and plants have travelled to their present forms as man has travelled to any one of his own most complicated inventions. Slowly, step by step, through many blunders and mischances which have worked together for good to those that have persevered in elasticity. They have travelled as man has travelled, with but little perception of a want till there was also some perception of a power, and with but little perception of a power till there was a dim sense of want; want stimulating power, and power stimulating want; and both so based upon each other that no one can say which is the true foundation, but rather that they must be both baseless and, as it were, meteoric in mid air. They have seen very little ahead of a present power or need, and have been then most moral, when most inclined to pierce a little into futurity, but also when most obstinately declining to pierce too far, and busy mainly with the present. They have been so far blindfolded that they could see but for a few steps in front of them, yet so far free to see that those steps were taken with aim and definitely, and not in the dark.

"Plus il a su," says Buffon, speaking of man, "plus il a pu, mais aussi moins il a fait, moins il a su." This holds good wherever life holds good. Wherever there is life there is a moral government of rewards and punishments understood by the amoeba neither better nor worse than by man. The history of organic development is the history of a moral struggle.

We know nothing as yet about the origin of a creature able to feel want and power, nor yet what want and power spring from. It does not seem worth while to go into these questions until an understanding has been come to as to whether the interaction of want and power in some low form or forms of life which could assimilate matter, reproduce themselves, vary their actions, and be capable of remembering, will or will not suffice to explain the development of the varied organs and desires which we see in the higher vertebrates and man. When this question has been settled, then it will be time to push our inquiries farther back.

But given such a low form of life as here postulated, and there is no force in Paley's pretended objection to the Darwinism of his time.

"Give our philosopher," he says, "appetencies; give him a portion of living irritable matter (a nerve or the clipping of a nerve) to work upon; give also to his incipient or progressive forms the power of propagating their like in every stage of their alteration; and if he is to be believed, he could replenish the world with all the vegetable and animal productions which we now see in it."[24]

After meeting this theory with answers which need not detain us, he continues:--

"The senses of animals appear to me quite incapable of receiving the explanation of their origin which this theory affords. Including under the word 'sense' the organ and the perception, we have no account of either. How will our philosopher get at vision or make an eye? Or, suppose the eye formed, would the perception follow? The same of the other senses. And this objection holds its force, ascribe what you will to the hand of time, to the power of habit, to changes too slow to be observed by man, or brought within any comparison which he is able to make of past things with the present. Concede what you please to these arbitrary and unattested superstitions, how will they help you? Here is no inception. No laws, no course, no powers of nature which prevail at present, nor any analogous to these would give commencement to a new sense; and it is in vain to inquire how that might proceed which would never *begin*."

In answer to this, let us suppose that some inhabitants of another world were to see a modern philosopher so using a microscope that they should believe it to be a part of the philosopher's own person, which he could cut off from and join again to himself at pleasure, and suppose there were a controversy as to how this microscope had originated, and that one party maintained the man had made it little by little because he wanted it, while the other declared this to be absurd and impossible; I ask, would this latter party be justified in arguing that microscopes could never have been perfected by degrees through the preservation of and accumulation of small successive improvements, inasmuch as men could not have begun to want to use microscopes until they had had a microscope which should show them that such an instrument would be useful to them, and that hence there is nothing to account for the *beginning* of microscopes, which might indeed make some progress when once originated, but which could never originate?

It might be pointed out to such a reasoner, firstly, that as regards any acquired power the various stages in the acquisition of which he might be supposed able to remember, he would find that, logic notwithstanding, the wish did originate the power, and yet was originated by it, both coming up gradually out of something which was not recognisable as either power or wish, and advancing through vain beating of the air, to a vague effort, and from this to definite effort with failure, and from this to definite effort with success, and from this to success with little consciousness of effort, and from this to success with such complete absence of effort that he now acts unconsciously and without power of introspection, and that, do what he will, he can rarely or never draw a sharp dividing line whereat anything shall be said to begin, though none less certain that there has been a continuity in discontinuity, and a discontinuity in continuity between it and certain other past things; moreover, that his opponents postulated so much beginning of the microscope as that there should be a dew drop, even as our evolutionists start with a sense of touch, of which sense all the others are modifications, so that not one of them but is resolvable into touch by more or less easy stages; and secondly, that the question is one of fact and of the more evident deductions therefrom, and should not be carried back to those remote beginnings where the nature of the facts is so purely a matter of conjecture and inference.

No plant or animal, then, according to our view, would be able to conceive more than a very slight improvement on its organization at a given time, so clearly as to make the efforts towards it that would result in growth of the required modification; nor would these efforts be made with any far-sighted perception of what next and next and after, but only of what next; while many of the happiest thoughts would come like all other happy thoughts--thoughtlessly; by a chain of reasoning too swift and subtle for conscious analysis by the individual, as will be more fully insisted on hereafter. Some of these modifications would be noticeable, but the majority would involve no more noticeable difference than can be detected between the length of the shortest day, and that of the shortest but one.

Thus a bird whose toes were not webbed, but who had under force of circumstances little by little in the course of many generations learned to swim, either from having lived near a lake, and having learnt the art owing to its fishing habits, or from wading about in shallow pools by the sea-side at low water, and finding itself sometimes a little out of its depth and just managing to scramble over the intermediate yard or so between it and safety--such a bird did not probably conceive the idea of swimming on the water and set itself to learn to do so, and then conceive the idea of webbed feet and set itself to get webbed feet. The bird found itself in some small difficulty, out of which it either saw, or at any rate found that it could extricate itself by striking out vigorously with its feet and extending its toes as far as ever it could; it thus began to learn the art of swimming and conceived the idea of swimming synchronously, or nearly so; or perhaps wishing to get over a yard or two of deep water, and trying to do so without being at the trouble of rising to fly, it would splash and struggle its way over the water, and thus practically swim, though without much perception of what it had been doing. Finding that no harm had come to it, the bird would do the same again, and again; it would thus presently lose fear, and would be able to act more calmly; then it would begin to find out that it could swim a little, and if its food lay much in the water so that it would be of great advantage to it to be able to alight and rest without being forced to return to land, it would begin to make a practice of swimming. It would now discover that it could swim the more easily according as its feet presented a more extended surface to the water; it would therefore keep its toes extended whenever it swam, and as far as in it lay, would make the most of whatever skin was already at the base of its toes. After very many generations it would become web-footed, if doing as above described should have been found continuously convenient, so that the bird should have continuously used the skin about its toes as much as possible in this direction.

For there is a margin in every organic structure (and perhaps more than we imagine in things inorganic also), which will admit of references, as it were, side notes, and glosses upon the original text. It is on this margin that we may err or wander--the greatness of a mistake depending rather upon the extent of the departure from the original text, than on the direction that the departure takes. A little error on the bad side is more pardonable, and less likely to hurt the organism than a too great departure upon the right one. This is a fundamental proposition in any true system of ethics, the question what is too much or too sudden being decided by much the same higgling as settles the price of butter in a country market, and being as invisible as the link which connects the last moment of desire with the first of power and performance, and with the material result achieved.

It is on this margin that the fulcrum is to be found, whereby we obtain the little purchase over our structure, that enables us to achieve great results if we use it steadily, with judgment, and with neither too little effort nor too much. It is by employing this that those who have a fancy to move their ears or toes without moving other organs learn to do so. There is a man at the Agricultural Hall now playing the violin with his toes, and playing it, as I am told, sufficiently well. The eye of the sailor, the wrist of the conjuror, the toe of the professional medium, are all found capable of development to an astonishing degree, even in a single lifetime; but in every case success has been attained by the simple process of making the best of whatever power a man has had at any given time, and by being on the look out to take advantage of accident, and even of misfortune. If a man would learn to paint, he must not theorize concerning art, nor think much what he would do beforehand, but he must do *something*--it does not matter what, except that it should be whatever at the moment will come handiest and easiest to him; and he must do that something as well as he can. This will presently open the door for something else, and a way will show itself which no conceivable amount of searching would have discovered, but which yet could never have been discovered by sitting still and taking no pains at all. "Dans l'animal," says Buffon, "il y a moins de jugement que de sentiment."[25]

It may appear as though this were blowing hot and cold with the same breath, inasmuch as I am insisting that important modifications of structure have been always purposive; and at the same time am denying that the creature modified has had any purpose in the greater part of all those actions which havc at length modified both structure and instinct. Thus I say that a bird learns to swim without having any purpose of learning to swim before it set itself to make those movements which have resulted in its being able to do so. At the same time I maintain that it has only learned to swim by trying to swim, and this involves the very purpose which I have just denied. The reconciliation of these two apparently irreconcilable contentions must be found in the consideration that the bird was not the less trying to swim, merely because it did not know the name we have chosen to give to the art which it was trying to master, nor yet how great were the resources of that art. A person, who knew all about swimming, if from some bank he could watch our supposed bird's first attempt to scramble over a short space of deep water, would at once declare that the bird was trying to swim--if not actually swimming. Provided then that there is a very little perception of, and prescience concerning, the means whereby the next desired end may be attained, it matters not how little in advance that end may be of present desires or faculties; it is still reached through purpose, and must be called purposive. Again, no matter how many of these small steps be taken, nor how absolute was the want of purpose or prescience concerning any but the one being actually taken at any given moment, this does not bar the result from having been arrived at through design and purpose. If each one of the small steps is purposive the result is purposive, though there was never purpose extended over more than one, two, or perhaps at most three, steps at a time.

Returning to the art of painting for an example, are we to say that the proficiency which such a student as was supposed above will certainly attain, is not due to design, merely because it was not until he had already become three parts excellent that he knew the full purport of all that he had been doing? When he began he had but vague notions of what he would do. He had a wish to learn to represent nature, but the line into which he has settled down has probably proved very different from that which he proposed to himself originally. Because he has taken advantage of his accidents, is it, therefore, one whit the less true that his success is the result of his desires and his design? The 'Times' pointed out not long ago that the theory which now associates meteors and comets in the most unmistakable manner, was suggested by one accident, and confirmed by another. But the writer added well that "such accidents happen only to the zealous student of nature's secrets." In the same way the bird that is taking to the habit of swimming, and of making the most of whatever skin it already has between its toes, will have doubtless to thank accidents for no small part of its progress; but they will be such accidents as could never have happened to, or been taken advantage of by any creature which was not zealously trying to make the most of itself--and between such accidents as this, and design, the line is hard to draw; for if we go deep enough we shall find that most of our design resolves itself into as it were a shaking of the bag to see what will come out that will suit our purpose, and yet at the same time that most of our shaking of the bag resolves itself into a design that the bag shall contain only such and such things, or thereabouts.

Again, the fact that animals are no longer conscious of design and purpose in much that they do, but act unreflectingly, and as we sometimes say concerning ourselves "automatically" or "mechanically"--that they have no idea whatever of the steps whereby they have travelled to their present state, and show no sign of doubt about what must have been at one time the subject of all manner of doubts, difficulties, and discussions--that whatever sign of reflection they now exhibit is to be found only in case of some novel feature or difficulty presenting itself; these facts do not bar that the results achieved should be attributed to an inception in reason, design, and purpose, no matter how rapidly and as we call it instinctively, the creatures may now act.

For if we look closely at such an invention as the steam engine in its latest and most complicated developments, about which there can be no dispute but that they are achievements of reason, purpose, and design, we shall find them present us with examples of all those features the presence of which in the handiwork of animals is too often held to bar reason and purpose from having had any share therein.

Assuredly such men as the Marquis of Worcester and Captain Savery had very imperfect ideas as to the upshot of their own action. The simplest steam engine now in use in England is probably a marvel of ingenuity as compared with the highest development which appeared possible to these two great men, while our newest and most highly complicated engines would seem to them more like living beings than machines. Many, again, of the steps leading to the present development have been due to action which had but little heed of the steam engine, being the inventions of attendants whose desire was to save themselves the trouble of turning this or that cock, and who were indifferent to any other end than their own immediate convenience. No step in fact along the whole route was ever taken with much perception of what would be the next step after the one being taken at any given moment.

Nor do we find that an engine made after any old and well-known pattern is now made with much more consciousness of design than we can suppose a bird's nest to be built with. The greater number of the parts of any such engine, are made by the gross as it were like screws and nuts, which are turned out by machinery and in respect of which the labour of design is now no more felt than is the design of him who first invented the wheel. It is only when circumstances require any modification in the article to be manufactured that thought and design will come into play again; but I take it few will deny that if circumstances compel a bird either to give up a nest three-parts built altogether, or to make some trifling deviation from its ordinary practice, it will in nine cases out of ten make such deviation as shall show that it had thought the matter over, and had on the whole concluded to take such and such a course, that is to say, that it had reasoned and had acted with such purpose as its reason had dictated.

And I imagine that this is the utmost that anyone can claim even for man's own boasted powers. Set the man who has been accustomed to make engines of one type, to make engines of another type without any intermediate course of training or instruction, and he will make no better figure with his engines than a thrush would do if commanded by her mate to make a nest like a blackbird. It is vain then to contend that the ease and certainty with which an action is performed, even though it may have now become matter of such fixed habit that it cannot be suddenly and seriously modified without rendering the whole performance abortive, is any argument against that action having been an achievement of design and reason in respect of each one of the steps that have led to it; and if in respect of each one of the steps then as regards the entire action; for we see our own most reasoned actions become no less easy, unerring, automatic, and unconscious, than the actions which we call instinctive when they have been repeated a sufficient number of times.

This has been often pointed out, but I insisted upon it and developed it in 'Life and Habit,' more I believe than has been done hitherto, at the same time making it the key to many phenomena of growth and heredity which without such key seem explained by words rather than by any corresponding peace of mind in our ideas concerning them. Seeing that I dwelt much on the importance of bearing in mind the vanishing tendency of consciousness, volition, and memory upon their becoming intense, a tendency which no one after five minutes' reflection will venture to deny, some reviewers have imagined that I am advocating the same views as have been put forward by Von Hartmann under the title of 'the Philosophy of the Unconscious.' Unless, however, I am much mistaken, their opinion is without foundation. For so far as I can gather, Von Hartmann personifies the unconscious and makes it act and think--in fact deifies it--whereas I only infer a certain history for certain of our growths and actions in consequence of observing that often repeated actions come in time to be performed unconsciously. I cannot think I have done more than note a fact which all must acknowledge, and drawn from it an inference which may or may not be true, but which is at any rate perfectly intelligible, whereas if Von Hartmann's meaning is anything like what Mr. Sully says it is,[26] I can only say that it has not been given to me to form any definite conception whatever as to what that meaning may be. I am encouraged moreover to hope that I am not in the same condemnation with Von Hartmann--if, indeed, Von Hartmann is to be condemned, about which I know nothing--by the following extract from a German Review of 'Life and Habit.'

"Der erste dieser beiden Erklärungsversuche, ist eine wahre 'Philosophie des Unbewussten' nicht des Hartmann'schen Unbewussten welches hellsehend und wunderthätig von aussen in die natürliche Entwickelung der Organismen eingreift, sondern eines Unbewussten welches wie der Verfasser zeigt, in allen organischen Wesen anzunehmen unsere eigene Erfahrung und die Stufenfolge der Organismen von den Moneren und Amoeben bis zu den höchsten Pflanzen und Thieren und uns selbst aufwärts--uns gestattet, wenn nicht uns nöthigt. Der Gedankengang dieser neuen oder wenigstens in diesem Sinne wohl zum ersten Male consequent im Einzelnen durchgeführten Philosophie des Unbewussten ist, seinen Hauptzügen nach kurz angedeutet, folgender."[27]

Even here I am made to personify more than I like; I do not wish to say that the unconscious does this or that, but that when we have done this or that sufficiently often we do it unconsciously.

If the foregoing be granted, and it be admitted that the unconsciousness and seeming automatism with which any action may be performed is no bar to its having a foundation in memory, reason, and at one time consciously recognized effort--and this I believe to be the chief addition which I have ventured to make to the theory of Buffon and Dr. Erasmus Darwin--then the wideness of the difference between the Darwinism of eighty years ago and the Darwinism of to-day becomes immediately apparent, and it also becomes apparent, how important and interesting is the issue which is raised between them.

According to the older Darwinism the lungs are just as purposive as the corkscrew. They, no less than the corkscrew, are a piece of mechanism designed and gradually improved upon and perfected by an intelligent creature for the gratification of its own needs. True there are many important differences between mechanism which is part of the body, and mechanism which is no such part, but the differences are such as do not affect the fact that in each case the result, whether, for example, lungs or corkscrew, is due to desire, invention, and design.

And now I will ask one more question, which may seem, perhaps, to have but little importance, but which I find personally interesting. I have been told by a reviewer, of whom upon the whole I have little reason to complain, that the theory I put forward in 'Life and Habit,' and which I am now again insisting on, is pessimism--pure and simple. I have a very vague idea what pessimism means, but I should be sorry to believe that I am a pessimist. Which, I would ask, is the pessimist? He who sees love of beauty, design, steadfastness of purpose, intelligence, courage, and every quality to which success has assigned the name of "worth," as having drawn the pattern of every leaf and organ now and in all past time, or he who sees nothing in the world of nature but a chapter of accidents and of forces interacting blindly?

FOOTNOTES:

[24] 'Nat. Theol.,' ch. xxiii.

[25] 'Oiseaux,' vol. i. p. 5.

[26] 'Westminster Review,' vol. xlix. p. 124.

[27] Translation: "The first of these two attempts is a true 'philosophy of the unconscious,' not Hartmann's unconscious, which influences the natural evolution of organism from without as though by Providence and miracle, but of an unconscious, which, as the author shows, our own experience and the progressive succession of organisms from the monads and amoebæ up to the highest plants and animals, including ourselves, allows, if it does not compel us to assume [as obtaining] in all organic beings. This philosophy of the unconscious is new, or at any rate now for the first time carried out consequentially in detail; its main features, briefly stated are as follows."

CHAPTER VI.

SCHEME OF THE REMAINDER OF THE WORK. HISTORICAL SKETCH OF THE THEORY OF EVOLUTION.

I have long felt that evolution must stand or fall according as it is made to rest or not on principles which shall give a definite purpose and direction to the variations whose accumulation results in specific, and ultimately in generic differences. In other words, according as it is made to stand upon the ground first clearly marked out for it by Dr. Erasmus Darwin and afterwards adopted by Lamarck, or on that taken by Mr. Charles Darwin.

There is some reason to fear that in consequence of the disfavour into which modern Darwinism is seen to be falling by those who are more closely watching the course of opinion upon this subject, evolution itself may be for a time discredited as something inseparable from the theory that it has come about mainly through "the means" of natural selection. If people are shown that the arguments by which a somewhat startling conclusion has been reached will not legitimately lead to that conclusion, they are very ready to assume that the conclusion must be altogether unfounded, especially when, as in the present case, there is a vast mass of vested interests opposed to the conclusion. Few know that there are other great works upon descent with modification besides Mr. Darwin's. Not one person in ten thousand has any distinct idea of what Buffon, Dr. Darwin, and Lamarck propounded. Their names have been discredited by the very authors who have been most indebted to them; there is hardly a writer on evolution who does not think it incumbent upon him to warn Lamarck off the ground which he at any rate made his own, and to cast a stone at what he will call the "shallow speculations" or "crude theories" or the "well-known doctrine" of the foremost exponent of Buffon and Dr. Darwin. Buffon is a great name, Dr. Darwin is no longer even this, and Lamarck has been so systematically laughed at that it amounts to little less than philosophical suicide for anyone to stand up in his behalf. Not one of our scientific elders or chief priests but would caution a student rather to avoid the three great men whom I have named than to consult them. It is a perilous task therefore to try and take evolution from the pedestal on which it now appears to stand so securely, and to put it back upon the one raised for it by its propounders; yet this is what I believe will have to be done sooner or later unless the now general acceptance of evolution is to be shaken more rudely than some of its upholders may anticipate. I propose therefore to give a short biographical sketch of the three writers whose works form new departures in the history of evolution, with a somewhat full *résumé* of the positions they took in regard to it. I will also touch briefly upon some other writers who have handled the same subject. The reader will thus be enabled to follow the development of a great conception as it has grown up in the minds of successive men of genius, and by thus growing with it, as it were, through its embryonic stages, he will make himself more thoroughly master of it in all its bearings.

I will then contrast the older with the newer Darwinism, and will show why the 'Origin of Species,' though an episode of incalculable value, cannot, any more than the 'Vestiges of Creation,' take permanent rank in the literature of evolution.

It will appear that the evolution of evolution has gone through the following principal stages:--

I. A general conception of the fact that specific types were not always immutable.

This was common to many writers, both ancient and modern; it has been occasionally asserted from the times of Anaximander and Lucretius to those of Bacon and Sir Walter Raleigh.

II. A definite conception that animal and vegetable forms were so extensively mutable that few (and, if so, perhaps but one) could claim to be of an original stock; the direct effect of changed conditions being assigned as the cause of modification, and the important consequences of the struggle for existence being in many respects fully recognized. The fact of design or purpose in connection with organism, as causing habits and thus as underlying all variation, was also indicated with some clearness, but was not thoroughly understood.

This phase must be identified with the name of Buffon, who, as I will show reason for believing, would have carried his theory much further if he had not felt that he had gone as far in the right direction as was then desirable. Buffon put forward his opinions, with great reserve and yet with hardly less frankness, in volume after volume from 1749 to 1788, the year of his death, but they do not appear to have taken root at once in France. They took root in England, and were thence transplanted back to France.

III. A development in England of the Buffonian system, marked by glimpses of the unity between offspring and parents, and broad suggestions to the effect that the former must be considered as capable of remembering, under certain circumstances, what had happened to it, and what it did, when it was part of the personality of those from whom it had descended.

A definite belief, openly expressed, that not only are many species mutable, but that all living forms, whether animal or vegetable, are descended from a single, or at any rate from not many, original low forms of life, and this as the direct consequence of the actions and requirements of the living forms themselves, and as the indirect consequence of changed conditions. A definite cause is thus supposed to underlie variations, and the resulting adaptations become purposive; but this was not said, nor, I am afraid, seen.

This is the original Darwinism of Dr. Erasmus Darwin. It was put forward in his 'Zoonomia,' in 1794, and was adopted almost in its entirety by Lamarck, who, when he had caught the leading idea (probably through a French translation of the 'Loves of the Plants,' which appeared in 1800), began to expound it in 1801; in 1802, 1803, 1806, and 1809, he developed it with greater fulness of detail than Dr. Darwin had done, but perhaps with a somewhat less nice sense of some important points. Till his death, in 1831, Lamarck, as far as age and blindness would permit, continued to devote himself to the exposition of the theory of descent with modification.

IV. A more distinct perception of the unity of parents and offspring, with a bolder reference of the facts of heredity (whether of structure or instinct), to memory pure and simple; a clearer perception of the consequences that follow from the survival of the fittest, and a just view of the relation in which those consequences stand to "the circumstance-suiting" power of animals and plants; a reference of the variations whose accumulation results in species, to the volition of the animal or plant which varies, and perhaps a dawning perception that all adaptations of structure to need must therefore be considered as "purposive."

This must be connected with Mr. Matthew's work on 'Naval Timber and Arboriculture,' which appeared in 1831. The remarks which it contains in reference to evolution are confined to an appendix, but when brought together, as by Mr. Matthew himself, in the 'Gardeners' Chronicle' for April 7, 1860, they form one of the most perfect yet succinct expositions of the theory of evolution that I have ever seen. I shall therefore give them in full.[28] This book was well received, and was reviewed in the 'Quarterly Review,'[29] but seems to have been valued rather for its views on naval timber than on evolution. Mr. Matthew's merit lies in a just appreciation of the importance of each one of the principal ideas which must be present in combination before we can have a correct conception of evolution, and of their bearings upon one another. In his scheme of evolution I find each part kept in due subordination to the others, so that the whole theory becomes more coherent and better articulated than I have elsewhere found it; but I do not detect any important addition to the ideas which Dr. Darwin and Lamarck had insisted upon.

I pass over the 'Vestiges of Creation,' which should be mentioned only as having, as Mr. Charles Darwin truly says, "done excellent service in this country, in calling attention to this subject, in removing prejudice, and in thus preparing the ground for the reception of analogous views."[30] The work neither made any addition to ideas which had been long familiar, nor arranged old ones in a satisfactory manner. Such as it is, it is Dr. Darwin and Lamarck, but Dr. Darwin and Lamarck spoiled. The first edition appeared in 1844.

I also pass over Isidore Geoffroy St. Hilaire's 'Natural History,' which appeared 1854-62, and the position of which is best described by calling it intermediate between the one which Buffon thought fit to pretend to take, and that actually taken by Lamarck. The same may be said also of Étienne Geoffroy. I will, however, just touch upon these writers later on.

A short notice, again, will suffice for the opinions of Goethe, Treviranus, and Oken, none of whom can I discover as having originated any important new idea; but knowing no German, I have taken this opinion from the résumé of each of these writers, given by Professor Haeckel in his 'History of Creation.'

V. A time of retrogression, during which we find but little apparent appreciation of the unity between parents and offspring; no reference to memory in connection with heredity, whether of instinct or structure; an exaggerated view of the consequences which may be deduced from the fact that the fittest commonly survive in the struggle for existence; the denial of any known principle as underlying variations; comparatively little appreciation of the circumstance-suiting power of plants and animals, and a rejection of purposiveness. By far the most important exponent of this phase of opinion concerning evolution is Mr. Charles Darwin, to whom, however, we are more deeply indebted than to any other living writer for the general acceptance of evolution in one shape or another. The 'Origin of Species' appeared in 1859, the same year, that is to say, as the second volume of Isidore Geoffroy's 'Histoire Naturelle Générale.'

VI. A reaction against modern Darwinism, with a demand for definite purpose and design as underlying variations. The best known writers who have taken this line are the Rev. J. J. Murphy and Professor Mivart, whose 'Habit and intelligence' and 'Genesis of Species' appeared in 1869 and 1871 respectively. In Germany Professor Hering has revived the idea of memory as explaining the phenomena of heredity satisfactorily, without probably having been more aware that it had been advanced already than I was myself when I put it forward recently in 'Life and Habit.' I have never seen the lecture in which Professor Hering has referred the phenomena of heredity to memory, but will give an extract from it which appeared in the 'Athenæum,' as translated by Professor Ray Lankester.[31] The only new feature which I believe I may claim to have added to received ideas concerning evolution, is a perception of the fact that the unconsciousness with which we go through our embryonic and infantile stages, and with which we discharge the greater number and more important of our natural functions, is of a piece with what we observe concerning all habitual actions, as well as concerning memory; an explanation of the phenomena of old age; and of the main principle which underlies longevity. I may, perhaps, claim also to have more fully explained the passage of reason into instinct than I yet know of its having been explained elsewhere.[32]

FOOTNOTES:

[28] See ch. xviii. of this volume.

[29] Vol. xlix. p. 125.

[30] 'Origin of Species,' Hist. Sketch, xvii.

[31] See page 199 of this volume.

[32] Apropos of this, a friend has kindly sent me the following extract from Balzac:--"Historiquement, les paysans sont encore au lendemain de la Jacquerie, leur défaite est restée inscrite dans leur cervelle. *Ils ne se souviennent plus du fait, il est passé à l'état d'idée instinctive.*"--Balzac, 'Les Paysans,' v.

CHAPTER VII.

PRE-BUFFONIAN EVOLUTION, AND SOME GERMAN WRITERS.

Let us now proceed to the fuller development of the foregoing sketch.

"Undoubtedly," says Isidore Geoffroy, "from the most ancient times many philosophers have imagined vaguely that one species can be transformed into another. This doctrine seems to have been adopted by the Ionian school from the sixth century before our era.... Undoubtedly also the same opinion reappeared on several occasions in the middle ages, and in modern times; it is to be found in some of the hermetic books, where the transmutation of animal and vegetable species, and that of metals, are treated as complementary to one another. In modern times we again find it alluded to by some philosophers, and especially by Bacon, whose boldness is on this point extreme. Admitting it as 'incontestable that plants sometimes degenerate so far as to become plants of another species,' Bacon did not hesitate to try and put his theory into practice. He tried, in 1635, to give 'the rules' for the art of changing 'plants of one species into those of another.'"

This must be an error. Bacon died in 1626. The passage of Bacon referred to is in 'Nat. Hist.,' Cent. vi. ("Experiments in consort touching the degenerating of plants, and the transmutation of them one into another"), and is as follows:--

"518. This rule is certain, that plants for want of culture degenerate to be baser in the same kind; and sometimes so far as to change into another kind. 1. The standing long and not being removed maketh them degenerate. 2. Drought unless the earth, of itself, be moist doth the like. 3. So doth removing into worse earth, or forbearing to compost the earth; as we see that water mint turneth into field mint, and the colewort into rape by neglect, &c."

"525. It is certain that in very steril years corn sown will grow to another kind:--

'Grandia sæpe quibus mandavimus hordea sulcis,
Infelix lolium, et steriles dominantur avenæ.'

And generally it is a rule that plants that are brought forth for culture, as corn, will sooner change into other species, than those that come of themselves; for that culture giveth but an adventitious nature, which is more easily put off."

Changed conditions, according to Bacon (though he does not use these words), appear to be "the first rule for the transmutation of plants."

"But how much value," continues M. Geoffroy, "ought to be attached to such prophetic glimpses, when they were neither led up to, nor justified by any serious study? They are conjectures only, which, while bearing evidence to the boldness or rashness of those who hazarded them, remain almost without effect upon the advance of science. Bacon excepted, they hardly deserve to be remembered. As for De Maillet, who makes birds spring from flying fishes, reptiles from creeping fishes, and men from tritons, his dreams, taken in part from Anaximander, should have their place not in the history of science, but in that of the aberrations of the human mind."[33]

A far more forcible and pregnant passage, however, is the following, from Sir Walter Raleigh's 'History of the World,' which Mr. Garnett has been good enough to point out to me:--

"For mine owne opinion I find no difference but only in magnitude between the Cat of Europe, and the Ounce of India; and even those dogges which are become wild in Hispagniola, with which the Spaniards used to devour the naked Indians, are now changed to Wolves, and begin to destroy the breed of their Cattell, and doe often times teare asunder their owne children. The common crow and rooke of India is full of red feathers in the droun'd and low islands of Caribana, and the blackbird and thrush hath his feathers mixt with black and carnation in the north parts of Virginia. The Dog-fish of England is the Sharke of the South Ocean. For if colour or magnitude made a difference of Species, then were the Negroes, which wee call the Blacke-Mores, *non animalia rationalia*, not Men but some kind of strange Beasts, and so the giants of the South America should be of another kind than the people of this part of the World. We also see it dayly that the nature of fruits are changed by transplantation."[34]

For information concerning the earliest German writers on evolution, I turn to Professor Haeckel's

'History of Creation,' and find Goethe's name to head the list. I do not gather, however, that Goethe added much to the ideas which Buffon had already made sufficiently familiar. Professor Haeckel does not seem to be aware of Buffon's work, and quotes Goethe as making an original discovery when he writes, in the year 1796:--"Thus much then we have gained, that we may assert without hesitation that all the more perfect organic natures, such as fishes, amphibious animals, birds, mammals, and man at the head of the last, were all formed upon one original type, which only varies more or less in parts which were none the less permanent, and still daily changes and modifies its form by propagation."[35] But these, as we shall see, are almost Buffon's own words--words too that Buffon insisted on for many years. Again Professor Haeckel quotes Goethe as writing in the year 1807:--

"If we consider plants and animals in their most imperfect condition, they can hardly be distinguished." This, however, had long been insisted upon by Bonnet and Dr. Erasmus Darwin, the first of whom was a naturalist of world-wide fame, while the 'Zoonomia' of Dr. Darwin had been translated into German between the years 1795 and 1797, and could hardly have been unknown to Goethe in 1807, who continues: "But this much we may say, that the creatures which by degrees emerge as plants and animals out of a common phase where they are barely distinguishable, arrive at perfection in two opposite directions, so that the plant in the end reaches its highest glory in the tree, which is immovable and stiff, the animal in man who possesses the greatest elasticity and freedom." Professor Haeckel considers this to be a remarkable passage, but I do not think it should cause its author to rank among the founders of the evolution theory, though he may justly claim to have been one of the first to adopt it. Goethe's anatomical researches appear to have been more important, but I cannot find that he insisted on any new principle, or grasped any unfamiliar conception, which had not been long since grasped and widely promulgated by Buffon and by Dr. Erasmus Darwin.

Treviranus (1776-1837), whom Professor Haeckel places second to Goethe, is clearly a disciple of Buffon, and uses the word "degeneration" in the same sense as Buffon used it many years earlier, that is to say, as "descent with modification," without any reference to whether the offspring was, as Buffon says, "perfectionné ou dégradé." He cannot claim, any more than Goethe, to rank as a principal figure in the history of evolution.

Of Oken, Professor Haeckel says that his 'Naturphilosophie,' which appeared in 1809--in the same year, that is to say, as the 'Philosophie Zoologique' of Lamarck--was "the nearest approach to the natural theory of descent, newly established by Mr. Charles Darwin," of any work that appeared in the first decade of our century. But I do not detect any important difference of principle between his system and that of Dr. Erasmus Darwin, among whose disciples he should be reckoned.

"We now turn," says Professor Haeckel after

referring to a few more German writers who adopted a belief in evolution, "from the German to the French nature-philosophers who have likewise held the theory of descent, since the beginning of this century. At their head stands Jean Lamarck, who occupies the first place next to Darwin and Goethe in the history of the doctrine of Filiation."[36] This is rather a surprising assertion, but I will leave the reader of the present volume to assign the value which should be attached to it.

Professor Haeckel devotes ten lines to Dr. Erasmus Darwin, who he declares "expresses views very similar to those of Goethe and Lamarck, without, however, *then* knowing anything about these two men;" which is all the more strange inasmuch as Dr. Darwin preceded them, and was a good deal better known to them, probably, than they to him; but it is plain Professor Haeckel has no acquaintance with the 'Zoonomia' of Dr. Erasmus Darwin. From all, then, that I am able to collect, I conclude that I shall best convey to the reader an idea of the different phases which the theory of descent with modification has gone through, by confining his attention almost entirely to Buffon, Dr. Erasmus Darwin, Lamarck, and Mr. Charles Darwin.

FOOTNOTES:

[33] 'Hist. Nat. Gen.,' vol. ii. p. 385, 1859.

[34] 'History of the World,' bk. i. ch. vii. § 9 ('Athenæum,' March 27, 1875).

[35] 'History of Creation,' vol. i. p. 91.

[36] 'History of Creation,' bk. i. ch. iii. (H. S. King, 1876).

CHAPTER VIII.

BUFFON--MEMOIR.

Buffon, says M. Flourens, was born at Montbar, on the 7th of September, 1707; he died in Paris, at the Jardin du Roi, on the 16th of April, 1788, aged 81 years. More than fifty of these years, as he used himself to say, he had passed at his writing-desk. His father was a councillor of the parliament of Burgundy. His mother was celebrated for her wit, and Buffon cherished her memory.

He studied at Dijon with much *éclat*, and shortly after leaving became accidentally acquainted with the Duke of Kingston, a young Englishman of his own age, who was travelling abroad with a tutor. The three travelled together in France and Italy, and Buffon then passed some months in England.

Returning to France, he translated Hales's 'Vegetable Statics' and Newton's 'Treatise on Fluxions.' He refers to several English writers on natural history in the course of his work, but I see he repeatedly spells the English name Willoughby, "Willulghby." He was appointed superintendent of the Jardin du Roi in 1739, and from thenceforth devoted himself to science.

In 1752 Buffon married Mdlle. de Saint Bélin, whose beauty and charm of manner were extolled by all her contemporaries. One son was born to him, who entered the army, became a colonel, and I grieve to say, was guillotined at the age of twenty-nine, a few days only before the extinction of the Reign of Terror.

Of this youth, who inherited the personal comeliness and ability of his father, little is recorded except the following story. Having fallen into the water and been nearly drowned when he was about twelve years old, he was afterwards accused of having been afraid: "I was so little afraid," he answered, "that though I had been offered the hundred years which my grandfather lived, I would have died then and there, if I could have added one year to the life of my father;" then thinking for a minute, a flush suffused his face, and he added, "but I should petition for one quarter of an hour in which to exult over the thought of what I was about to do."

On the scaffold he showed much composure, smiling half proudly, half reproachfully, yet wholly kindly upon the crowd in front of him. "Citoyens," he said, "Je me nomme Buffon," and laid his head upon the block.

The noblest outcome of the old and decaying order, overwhelmed in the most hateful birth frenzy of the new. So in those cataclysms and revolutions which take place in our own bodies during their development, when we seem studying in order to become fishes and suddenly make, as it were, different arrangements and resolve on becoming men--so, doubtless, many good cells must go, and their united death cry comes up, it may be, in the pain which an infant feels on teething.

But to return. The man who could be father of such a son, and who could retain that son's affection, as it is well known that Buffon retained it, may not perhaps always be strictly accurate, but it will be as well to pay attention to whatever he may think fit to tell us. These are the only people whom it is worth while to look to and study from.

"Glory," said Buffon, after speaking of the hours during which he had laboured, "glory comes always after labour if she can--*and she generally can.*" But in his case she could not well help herself. "He was conspicuous," says M. Flourens, "for elevation and force of character, for a love of greatness and true magnificence in all he did. His great wealth, his handsome person, and graceful manners seemed in correspondence with the splendour of his genius, so that of all the gifts which Fortune has it in her power to bestow she had denied him nothing."

Many of his epigrammatic sayings have passed into proverbs: for example, that "genius is but a supreme capacity for taking pains." Another and still more celebrated passage shall be given in its entirety and with its original setting.

"Style," says Buffon, "is the only passport to posterity. It is not range of information, nor mastery of some little known branch of science, nor yet novelty of matter that will ensure immortality.

Works that can claim all this will yet die if they are conversant about trivial objects only, or written without taste, genius and true nobility of mind; for range of information, knowledge of details, novelty of discovery are of a volatile essence and fly off readily into other hands that know better how to treat them. The matter is foreign to the man, and is not of him; the manner is the man himself."[37]

"Le style, c'est l'homme même." Elsewhere he tells us what true style is, but I quote from memory and cannot be sure of the passage. "Le style," he says, "est comme le bonheur; il vient de la douceur de l'âme."

Is it possible not to think of the following?--

"But whether there be prophecies they shall fail; whether there be tongues they shall cease; whether there be knowledge it shall vanish away ... and now abideth faith, hope and charity, these three; but the greatest of these is charity."[38]

FOOTNOTES:

[37] 'Discours de Réception à l'Académie Française.'

[38] 1 Cor. xiii. 8, 13.

CHAPTER IX.

BUFFON'S METHOD--THE IRONICAL CHARACTER OF HIS WORK.

Buffon's idea of a method amounts almost to the denial of the possibility of method at all. "The true method," he writes, "is the complete description and exact history of each particular object,"[39] and later on he asks, "is it not more simple, more natural and more true to call an ass an ass, and a cat a cat, than to say, without knowing why, that an ass is a horse, and a cat a lynx."[40]

He admits such divisions as between animals and vegetables, or between vegetables and minerals, but that done, he rejects all others that can be founded on the nature of things themselves. He concludes that one who could see things in their entirety and without preconceived opinions, would classify animals according to the relations in which he found himself standing towards them:--

"Those which he finds most necessary and useful to him will occupy the first rank; thus he will give the precedence among the lower animals to the dog and the horse; he will next concern himself with those which without being domesticated, nevertheless occupy the same country and climate as himself, as for example stags, hares, and all wild animals; nor will it be till after he has familiarized himself with all these that curiosity will lead him to inquire what inhabitants there may be in foreign climates, such as elephants, dromedaries, &c. The same will hold good for fishes, birds, insects, shells, and for all nature's other productions; he will study them in proportion to the profit which he can draw from them; he will consider them in that order in which they enter into his daily life; he will arrange them in his head according to this order, which is in fact that in which he has become acquainted with them, and in which it concerns him to think about them. This order--the most natural of all--is the one which I have thought it well to follow in this volume. My classification has no more mystery in it than the reader has just seen ... it is preferable to the most profound and ingenious that can be conceived, for there is none of all the classifications which ever have been made or ever can be, which has not more of an arbitrary character than this has. Take it for all in all," he concludes, "it is more easy, more agreeable, and more useful, to consider things in their relation to ourselves than from any other standpoint."[41]

"Has it not a better effect not only in a treatise on natural history, but in a picture or any work of art to arrange objects in the order and place in which they are commonly found, than to force them into association in virtue of some theory of our own? Is it not better to let the dog which has toes, come after the horse which has a single hoof, in the same way as we see him follow the horse in daily life, than to follow up the horse by the zebra, an animal which is little known to us, and which has no other connection with the horse than the fact that it has a single hoof?"[42]

Can we suppose that Buffon really saw no more connection than this? The writer whom we shall presently find[43] declining to admit any essential difference between the skeletons of man and of the horse, can here see no resemblance between the zebra and the horse, except that they each have a single hoof. Is he to be taken at his word?

It is perhaps necessary to tell the reader that Buffon carried the foregoing scheme into practice as nearly as he could in the first fifteen volumes of his 'Natural History.' He begins with man--and then goes on to the horse, the ass, the cow, sheep, goat, pig, dog, &c. One would be glad to know whether he found it always more easy to decide in what order of familiarity this or that animal would stand to the majority of his readers than other classifiers have found it to know whether an individual more resembles one species or another; probably he never gave the matter a thought after he had gone through the first dozen most familiar animals, but settled generally down into a classification which becomes more and more specific--as when he treats of the apes and monkeys--till he reaches the birds, when he openly abandons his original idea, in deference, as he says, to the opinion of "le peuple des naturalistes."

Perhaps the key to this piece of apparent extravagance is to be found in the word "mystérieuse."[44] Buffon wished to raise a standing protest against mystery mongering. Or perhaps more probably, he wished at once "to turn to animals and plants under domestication," so as to insist early on the main object of his work--the plasticity of animal forms.

I am inclined to think that a vein of irony pervades the whole, or much the greater part of Buffon's work, and that he intended to convey, one meaning to one set of readers, and another to another; indeed, it is often impossible to believe that he is not writing between his lines for the discerning, what the undiscerning were not intended to see. It must be remembered that his 'Natural History' has two sides,--a scientific and a popular one. May we not imagine that Buffon would be unwilling to debar himself from speaking to those who could understand him, and yet would wish like Handel and Shakespeare to address the many, as well as the few? But the only manner in which these seemingly irreconcilable ends could be attained, would be by the use of language which should be self-adjusting to the capacity of the reader. So keen an observer can hardly have been blind to the signs of the times which were already close at hand. Free-thinker though he was, he was also a powerful member of the aristocracy, and little likely to demean himself--for so he would doubtless hold it--by playing the part of Voltaire or Rousseau. He would help those who could see to see still further, but he would not dazzle eyes that were yet imperfect with a light brighter than they could stand. He would therefore impose upon people, as much as he thought was for their good; but, on the other hand, he would not allow inferior men to mystify them.

"In the private character of Buffon," says Sir William Jardine in a characteristic passage, "we regret there is not much to praise; his disposition was kind and benevolent, and he was generally beloved by his inferiors, followers, and dependents, which were numerous over his extensive property; he was strictly honourable, and was an affectionate parent. In early youth he had entered into the pleasures and dissipations of life, and licentious habits seem to have been retained to the end. But the great blemish in such a mind was his declared infidelity; it presents one of those exceptions among the persons who have been devoted to the study of nature; and it is not easy to imagine a mind apparently with such powers, scarcely acknowledging a Creator, and when noticed, only by an arraignment for what appeared wanting or defective in his great works. So openly, indeed, was the freedom of his religious opinions expressed, that the indignation of the Sorbonne was provoked. He had to enter into an explanation which he in some way rendered satisfactory; and while he afterwards attended to the outward ordinances of religion, he considered them as a system of faith for the multitude, and regarded those most impolitic who most opposed them."[45]

This is partly correct and partly not. Buffon was a free-thinker, and as I have sufficiently explained, a decided opponent of the doctrine that rudimentary and therefore useless organs were designed by a Creator in order to serve some useful end throughout all time to the creature in which they are found.

He was not, surely, to hide the magnificent conceptions which he had been the first to grasp, from those who were worthy to receive them; on the other hand he would not tell the uninstructed what they would interpret as a license to do whatever they pleased, inasmuch as there was no God. What he did was to point so irresistibly in the right direction, that a reader of any intelligence should be in no doubt as to the road he ought to take, and then to contradict himself so flatly as to reassure those who would be shocked by a truth for which they were not yet ready. If I am right in the view which I have taken of Buffon's work, it is not easy to see how he could have formed a finer scheme, nor have carried it out more finely.

I should, however, warn the reader to be on his guard against accepting my view too hastily. So far as I know I stand alone in taking it. Neither Dr. Darwin nor Flourens, nor Isidore Geoffroy, nor Mr. Charles Darwin see any subrisive humour in Buffon's pages; but it must be remembered that Flourens was a strong opponent of mutability, and probably paid but little heed to what Buffon said on this question; Isidore Geoffroy is not a safe guide, as will appear presently; Mr. Charles Darwin seems to have adopted the one half of Isidore Geoffroy's conclusions without verifying either; and Dr. Erasmus Darwin, who has no small share of a very pleasant conscious humour, yet sometimes rises to such heights of unconscious humour, that Buffon's puny labour may well have been invisible to him. Dr. Darwin wrote a great deal of poetry, some of which was about the common pump. Miss Seward tells us, as we shall see later on, that he "illustrated this familiar object with a picture of Maternal Beauty administering sustenance to her infant." Buffon could not have done anything like this.

Buffon never, then, "arraigned the Creator for what

was wanting or defective in His works;" on the contrary, whenever he has led up by an irresistible chain of reasoning to conclusions which should make men recast their ideas concerning the Deity, he invariably retreats under cover of an appeal to revelation. Naturally enough, the Sorbonne objected to an artifice which even Buffon could not conceal completely. They did not like being undermined; like Buffon himself, they preferred imposing upon the people, to seeing others do so. Buffon made his peace with the Sorbonne immediately, and, perhaps, from that time forward, contradicted himself a little more impudently than heretofore.

It is probably for the reasons above suggested that Buffon did not propound a connected scheme of evolution or descent with modification, but scattered his theory in fragments up and down his work in the prefatory remarks with which he introduces the more striking animals or classes of animals. He never wastes evolutionary matter in the preface to an uninteresting animal; and the more interesting the animal, the more evolution will there be commonly found. When he comes to describe the animal more familiarly--and he generally begins a fresh chapter or half chapter when he does so--he writes no more about evolution, but gives an admirable description, which no one can fail to enjoy, and which I cannot think is nearly so inaccurate as is commonly supposed. These descriptions are the parts which Buffon intended for the general reader, expecting, doubtless, and desiring that such a reader should skip the dry parts he had been addressing to the more studious. It is true the descriptions are written *ad captandum*, as are all great works, but they succeed in captivating, having been composed with all the pains a man of genius and of great perseverance could bestow upon them. If I am not mistaken, he looked to these parts of his work to keep the whole alive till the time should come when the philosophical side of his writings should be understood and appreciated.

Thus the goat breeds with the sheep, and may therefore serve as the text for a dissertation on hybridism, which is accordingly given in the preface to this animal. The presence of rudimentary organs under a pig's hoof suggests an attack upon the doctrine of final causes in so far as it is pretended that every part of every animal or plant was specially designed with a view to the wants of the animal or plant itself once and for ever throughout all time. The dog with his great variety of breeds gives an opportunity for an article on the formation of breeds and sub-breeds by man's artificial selection. The cat is not honoured with any philosophical reflections, and comes in for nothing but abuse. The hare suggests the rabbit, and the rabbit is a rapid breeder, although the hare is an unusually slow one; but this is near enough, so the hare shall serve us for the theme of a discourse on the geometrical ratio of increase and the balance of power which may be observed in nature. When we come to the carnivora, additional reflections follow upon the necessity for death, and even for violent death; this leads to the question whether the creatures that are killed suffer pain; here, then, will be the proper place for considering the sensations of animals generally.

Perhaps the most pregnant passage concerning evolution is to be found in the preface to the ass, which is so near the beginning of the work as to be only the second animal of which Buffon treats after having described man himself. It points strongly in the direction of his having believed all animal forms to have been descended from one single common ancestral type. Buffon did not probably choose to take his very first opportunity in order to insist upon matter that should point in this direction; but the considerations were too important to be deferred long, and are accordingly put forward under cover of the ass, his second animal.

When we consider the force with which Buffon's conclusion is led up to; the obviousness of the conclusion itself when the premises are once admitted; the impossibility that such a conclusion should be again lost sight of if the reasonableness of its being drawn had been once admitted; the position in his scheme which is assigned to it by its propounder; the persistency with which he demonstrates during forty years thereafter that the premises, which he has declared should establish the conclusion in question, are indisputable;--when we consider, too, that we are dealing with a man of unquestionable genius, and that the times and circumstances of his life were such as would go far to explain reserve and irony--is it, I would ask, reasonable to suppose that Buffon did not, in his own mind, and from the first, draw the inference to which he leads his reader, merely because from time to time he tells the reader, with a shrug of the shoulders, that *he* draws no inferences opposed to the Book of Genesis? Is it not more likely that

Buffon intended his reader to draw his inferences for himself, and perhaps to value them all the more highly on that account?

The passage to which I am alluding is as follows:--

"If from the boundless variety which animated nature presents to us, we choose the body of some animal or even that of man himself to serve as a model with which to compare the bodies of other organized beings, we shall find that though all these beings have an individuality of their own, and are distinguished from one another by differences of which the gradations are infinitely subtle, there exists at the same time a primitive and general design which we can follow for a long way, and the departures from which (*dégénérations*) are far more gentle than those from mere outward resemblance. For not to mention organs of digestion, circulation, and generation, which are common to all animals, and without which the animal would cease to be an animal, and could neither continue to exist nor reproduce itself--there is none the less even in those very parts which constitute the main difference in outward appearance, a striking resemblance which carries with it irresistibly the idea of a single pattern after which all would appear to have been conceived. The horse, for example--what can at first sight seem more unlike mankind? Yet when we compare man and horse point by point and detail by detail, is not our wonder excited rather by the points of resemblance than of difference that are to be found between them? Take the skeleton of a man; bend forward the bones in the region of the pelvis, shorten the thigh bones, and those of the leg and arm, lengthen those of the feet and hands, run the joints together, lengthen the jaws, and shorten the frontal bone, finally, lengthen the spine, and the skeleton will now be that of a man no longer, but will have become that of a horse--for it is easy to imagine that in lengthening the spine and the jaws we shall at the same time have increased the number of the vertebræ, ribs, and teeth. It is but in the number of these bones, which may be considered accessory, and by the lengthening, shortening, or mode of attachment of others, that the skeleton of the horse differs from that of the human body.... We find ribs in man, in all the quadrupeds, in birds, in fishes, and we may find traces of them as far down as the turtle, in which they seem still to be sketched out by means of furrows that are to be found beneath the shell. Let it be remembered that the foot of the horse, which seems so different from a man's hand, is, nevertheless, as M. Daubenton has pointed out, composed of the same bones, and that we have at the end of each of our fingers a nail corresponding to the hoof of a horse's foot. Judge, then, whether this hidden resemblance is not more marvellous than any outward differences--whether this constancy to a single plan of structure which we may follow from man to the quadrupeds, from the quadrupeds to the cetacea, from the cetacea to birds, from birds to reptiles, from reptiles to fishes--in which all such essential parts as heart, intestines, spine, are invariably found--whether, I say, this does not seem to indicate that the Creator when He made them would use but a single main idea, though at the same time varying it in every conceivable way, so that man might admire equally the magnificence of the execution and the simplicity of the design.[46]

"If we regard the matter thus, not only the ass and the horse, *but even man himself, the apes, the quadrupeds, and all animals might be regarded but as forming members of one and the same family*. But are we to conclude that within this vast family which the Creator has called into existence out of nothing, there are other and smaller families, projected as it were by Nature, and brought forth by her in the natural course of events and after a long time, of which some contain but two members, as the ass and the horse, others many members, as the weasel, martin, stoat, ferret, &c., and that on the same principle there are families of vegetables, containing ten, twenty, or thirty plants, as the case may be? If such families had any real existence they could have been formed only by crossing, by the accumulation of successive variations (*variation successive*), and by degeneration from an original type; but if we once admit that there are families of plants and animals, so that the ass may be of the family of the horse, and that the one may only differ from the other through degeneration from a common ancestor, we might be driven to admit that the ape is of the family of man, that he is but a degenerate man, and that he and man have had a common ancestor, even as the ass and horse have had. It would follow then that every family, whether animal or vegetable, had sprung from a single stock, which after a succession of generations, had become higher in the case of some of its descendants and lower in that of others."

What inference could be more aptly drawn? But it was not one which Buffon was going to put before the general public. He had said enough for the discerning, and continues with what is intended to make the conclusions they should draw even plainer to them, while it conceals them still more carefully from the general reader.

"The naturalists who are so ready to establish families among animals and vegetables, do not seem to have sufficiently considered the consequences which should follow from their premises, for these would limit direct creation to as small a number of forms as anyone might think fit (reduisoient le produit immédiat de la création, à un nombre d'individus aussi petit que l'on voudroit). *For if it were once shown that we had right grounds for establishing these families; if the point were once gained that among animals and vegetables there had been, I do not say several species, but even a single one, which had been produced in the course of direct descent from another species; if for example it could be once shown that the ass was but a degeneration from the horse--then there is no further limit to be set to the power of nature, and we should not be wrong in supposing that with sufficient time she could have evolved all other organized forms from one primordial type (et l'on n'auroit pas tort de supposer, que d'un seul être elle a su tirer avec le temps tous les autres êtres organisés).*"

Buffon now felt that he had sailed as near the wind as was desirable. His next sentence is as follows:--

"But no! It is certain *from revelation* that all animals have alike been favoured with the grace of an act of direct creation, and that the first pair of every species issued full formed from the hands of the Creator."[47]

This might be taken as *bonâ fide*, if it had been written by Bonnet, but it is impossible to accept it from Buffon. It is only those who judge him at second hand, or by isolated passages, who can hold that he failed to see the consequences of his own premises. No one could have seen more clearly, nor have said more lucidly, what should suffice to show a sympathetic reader the conclusion he ought to come to. Even when ironical, his irony is not the ill-natured irony of one who is merely amusing himself at other people's expense, but the serious and legitimate irony of one who must either limit the circle of those to whom he appeals, or must know how to make the same language appeal differently to the different capacities of his readers, and who trusts to the good sense of the discerning to understand the difficulty of his position, and make due allowance for it.

dog whistle

The compromise which he thought fit to put before the public was that "Each species has a type of which the principal features are engraved in indelible and eternally permanent characters, while all accessory touches vary."[48] It would be satisfactory to know where an accessory touch is supposed to begin and end.

!?

And again:--

"The essential characteristics of every animal have been conserved without alteration in their most important parts.... The individuals of each genus still represent the same forms as they did in the earliest ages, especially in the case of the larger animals" (so that the generic forms even of the larger animals prove not to be the same, but only 'especially' the same as in the earliest ages).[49]

This transparently illogical position is maintained ostensibly from first to last, much in the same spirit as in the two foregoing passages, written at intervals of thirteen years. But they are to be read by the light of the earlier one--placed as a lantern to the wary upon the threshold of his work in 1753--to the effect that a single, well substantiated case of degeneration would make it conceivable that all living beings were descended from a single common ancestor. If after having led up to this by a remorseless logic, a man is found five-and-twenty years later still substantiating cases of degeneration, as he has been substantiating them unceasingly in thirty quartos during the whole interval, there should be little question how seriously we are to take him when he wishes us to stop short of the conclusions he has told us we ought to draw from the premises that he has made it the business of his life to establish--especially when we know that he has a Sorbonne to keep a sharp eye upon him.

I believe that if the reader will bear in mind the twofold, serious and ironical, character of Buffon's work he will understand it, and feel an admiration for it which will grow continually greater and greater the more he studies it, otherwise he will miss the whole point.

Buffon on one of the early pages of his first volume protested against the introduction of either "*plaisanterie*" or "*équivoque*" (p. 25) into a serious work. But I have observed that there is an unconscious irony in most disclaimers of this nature. When a writer begins by saying that he has "an ineradicable tendency to make things clear," we may infer that we are going to be puzzled; so when he shows that he is haunted by a sense of the impropriety of allowing humour to intrude into his work, we may hope to be amused as well as interested. As showing how far the objection to humour which he expressed upon his twenty-fifth page succeeded in carrying him safely over his twenty-sixth and twenty-seventh, I will quote the following, which begins on page twenty-six:--

"Aldrovandus is the most learned and laborious of all naturalists; after sixty years of work he has left an immense number of volumes behind him, which have been printed at various times, the greater number of them after his death. It would be possible to reduce them to a tenth part if we could rid them of all useless and foreign matter, and of a prolixity which I find almost overwhelming; were this only done, his books should be regarded as among the best we have on the subject of natural history in its entirety. The plan of his work is good, his classification distinguished for its good sense, his dividing lines well marked, his descriptions sufficiently accurate--monotonous it is true, but painstaking; the historical part of his work is less good; it is often confused and fabulous, and the author shows too manifestly the credulous tendencies of his mind.

"While going over his work, I have been struck with that defect, or rather excess, which we find in almost all the books of a hundred or a couple of hundred years ago, and which prevails still among the Germans--I mean with that quantity of useless erudition with which they intentionally swell out their works, and the result of which is that their subject is overlaid with a mass of extraneous matter on which they enlarge with great complacency, but with no consideration whatever for their readers. They seem, in fact, to have forgotten what they have to say in their endeavour to tell us what has been said by other people.

"I picture to myself a man like Aldrovandus, after he has once conceived the design of writing a complete natural history. I see him in his library reading, one after the other, ancients, moderns, philosophers, theologians, jurisconsults, historians, travellers, poets, and reading with no other end than with that of catching at all words and phrases which can be forced from far or near into some kind of relation with his subject. I see him copying all these passages, or getting them copied for him, and arranging them in alphabetical order. He fills many portfolios with all manner of notes, often taken without either discrimination or research, and at last sets himself to write with a resolve that not one of all these notes shall remain unused. The result is that when he comes to his account of the cow or of the hen, he will tell us all that has ever yet been said about cows or hens; all that the ancients ever thought about them; all that has ever been imagined concerning their virtues, characters, and courage; every purpose to which they have ever yet been put; every story of every old woman that he can lay hold of; all the miracles which certain religions have ascribed to them; all the superstitions they have given rise to; all the metaphors and allegories which poets have drawn from them; the attributes that have been assigned to them; the representations that have been made of them in hieroglyphics and armorial bearings, in a word all the histories and all fables in which there was ever yet any mention either of a cow or hen. How much natural history is likely to be found in such a lumber room? and how is one to lay one's hand upon the little that there may actually be?"[50]

It is hoped that the reader will see Buffon, much us Buffon saw the learned Aldrovandus. He should see him going into his library, &c., and quietly chuckling to himself as he wrote such a passage as the one in which we lately found him saying that the larger animals had "especially" the same generic forms as they had always had. And the reader should probably see Daubenton chuckling also.

FOOTNOTES:

[39] Tom. i. p. 24, 1749.

[40] Tom. i. p. 40, 1749.

[41] Vol. i. p. 34, 1749.

[42] Tom. i. p. 36.

[43] See p. 88 of this volume; see also p. 155, and 164.

[44] Tom. i. p. 33.

[45] 'The Naturalist's Library,' vol. ii. p. 23, Edinburgh, 1843.

[46] Tom. iv. p. 381, 1753.

[47] Tom. iv. p. 383, 1753 (this was the first volume on the lower animals).

[48] Tom. xiii. p. ix. 1765.

[49] Sup. tom. v. p. 27, 1778.

[50] Tom. i. p. 28, 1749.

CHAPTER X.

SUPPOSED FLUCTUATIONS OF OPINION--CAUSES OR MEANS OF THE TRANSFORMATION OF SPECIES.

Enough, perhaps, has been already said to disabuse the reader's mind of the common misconception of Buffon, namely, that he was more or less of an elegant trifler with science, who cared rather about the language in which his ideas were clothed than about the ideas themselves, and that he did not hold the same opinions for long together; but the accusation of instability has been made in such high quarters that it is necessary to refute it still more completely.

Mr. Darwin, for example, in his "Historical Sketch of the Recent Progress of Opinion on the Origin of Species" prefixed to all the later editions of his own 'Origin of Species,' says of Buffon that he "was the first author who, in modern times, has treated" the origin of species "in a scientific spirit. But," he continues, "as his opinions fluctuated greatly at different periods, and as he does not enter on the causes or means of the transformation of species, I need not here enter on details."[51]

Mr. Darwin seems to have followed the one half of Isidore Geoffroy St. Hilaire's "full account of Buffon's conclusions" upon the subject of descent with modification,[52] to which he refers with approval on the second page of his historical sketch.[53]

Turning, then, to Isidore Geoffroy's work, I find that in like manner he too has been following the one half of what Buffon actually said. But even so, he awards Buffon very high praise.

"Buffon," he writes, "is to the doctrine of the mutability of species what Linnæus is to that of its fixity. It is only since the appearance of Buffon's 'Natural History,' and in consequence thereof, that the mutability of species has taken rank among scientific questions."[54]

"Buffon, who comes next in chronological order after Bacon, follows him in no other respect than that of time. He is entirely original in arriving at the doctrine of the variability of organic types, and in enouncing it after long hesitation, during which one can watch the labour of a great intelligence freeing itself little by little from the yoke of orthodoxy.

"But from this source come difficulties in the interpretation of Buffon's work which have misled many writers. Buffon expresses absolutely different opinions in different parts of his natural history--so much so that partisans and opponents of the doctrine of the fixity of species have alike believed and still believe themselves at liberty to claim Buffon as one of the great authorities upon their side."

Then follow the quotations upon which M. Geoffroy relies--to which I will return presently--after which the conclusion runs thus:--

"The dates, however, of the several passages in question are sufficient to explain the differences in their tenor, in a manner worthy of Buffon. Where are the passages in which Buffon affirms the immutability of species? At the beginning of his work. His first volume on animals[55] is dated 1753. The two following are those in which Buffon still shares the views of Linnæus; they are dated 1755 and 1756. Of what date are those in which Buffon declares for variability? From 1761 to 1766. And those in which, after having admitted variability and declared in favour of it, he proceeds to limit it? From 1765 to 1778.

"The inference is sufficiently simple. Buffon does but correct himself. He does not fluctuate. He goes once for all from one opinion to the other, from what he accepted at starting on the authority of another to what he recognized as true after twenty years of research. If while trying to set himself free from the prevailing notions, he in the first instance went, like all other innovators, somewhat to the opposite extreme, he essays as soon as may be to retrace his steps in some measure, and thenceforward to remain unchanged.

"Let the reader cast his eye over the general table of contents wherein Buffon, at the end of his 'Natural History,' gives a *résumé* of all of it that he is anxious to preserve. He passes over alike the passages in which he affirms and those in which he unreservedly denies the immutability of species, and indicates only the doctrine of the permanence of essential features and the variability of details (toutes les touches accessoires); he repeats this eleven years later in his 'Époques de la Nature'" (published 1778).[56]

But I think I can show that the passages which M. Geoffroy brings forward, to prove that Buffon was in the first instance a supporter of invariability, do not bear him out in the deduction he has endeavoured to draw from them.

"What author," he asks, "has ever pronounced more decidedly than Buffon in favour of the invariability of species? Where can we find a more decided expression of opinion than the following?

"'The different species of animals are separated from one another by a space which Nature cannot overstep.'"

On turning, however, to Buffon himself, I find the passage to stand as follows:--

"*Although* the different species of animals are separated from one another by a space which Nature cannot overstep--*yet some of them approach so nearly to one another in so many respects that there is only room enough left for the getting in of a line of separation between them*,"[57] and on the following page he distinctly encourages the idea of the mutability of species in the following passage:--

"In place of regarding the ass as a degenerate horse, there would be more reason in calling the horse a more perfect kind of ass (un âne perfectionné), and the sheep a more delicate kind of goat, that we have tended, perfected, and propagated for our use, and that the more perfect animals in general--especially the domestic animals--*draw their origin from some less perfect species of that kind of wild animal which they most resemble. Nature alone not being able to do as much as Nature and man can do in concert with one another*."[58]

But Buffon had long ago declared that if the horse and the ass could be considered as being blood relations there was no stopping short of the admission that all animals might also be blood relations--that is to say, descended from common ancestors--and now he tells us that the ass and horse *are* in all probability descended from common ancestors. Will a reader of any literary experience hold that so laborious, and yet so witty a writer, and one so studious of artistic effect, could ignore the broad lines he had laid down for himself, or forget how what he had said would bear on subsequent passages, and subsequent passages on it? A less painstaking author than Buffon may yet be trusted to remember his own work well enough to avoid such literary bad workmanship as this. If Buffon had seen reason to change his mind he would have said so, and would have contradicted the inference he had originally pronounced to be deducible from an admission of kinship between the ass and the horse. This, it is hardly necessary to say, he never does, though he frequently thinks it well to remind his reader of the fact that the ass and the horse are in all probability closely related. This is bringing two and two together with sufficient closeness for all practical purposes.

Should not M. Geoffroy's question, then, have rather been "Who has ever pronounced more grudgingly, even in an early volume, &c., &c., and who has more completely neutralized whatever concession he might appear to have been making?"

Nor does the only other passage which M. Geoffroy brings forward to prove that Buffon was originally a believer in the fixity of species bear him out much better. It is to be found on the opening page of a brief introduction to the wild animals. M. Geoffroy quotes it thus: "We shall see Nature dictating her laws, so simple yet so unchangeable, and imprinting her own immutable characters upon every species." But M. Geoffroy does not give the passage which, on the same page, admits mutability among domesticated animals, in the case of which he declares we find Nature "rarement perfectionnée, souvent alterée, défigurée;" nor yet does he deem it necessary to show that the context proves that this unchangeableness of wild animals is only relative; and this he should certainly have done, for two pages later on Buffon speaks of the American tigers, lions, and panthers as being "degenerated, if their original nature was cruel and ferocious; or, rather, they have experienced the effect of climate, and under a milder sky have assumed a milder nature, their excesses have become moderated, and by the changes which they have undergone they have become more in conformity with the country they inhabit."[59]

And again:--

"If we consider each species in the different climates which it inhabits, we shall find perceptible varieties as regards size and form: they all derive an impress to a greater or less extent from the climate in which they live. *These changes are only made slowly and imperceptibly.* Nature's great workman is Time. He marches ever with an even pace, and does nothing by leaps and bounds, but by degrees, gradations, and succession he does all things; and the changes which he works--at first imperceptible--become little by little perceptible, and show themselves eventually in results about which there can be no mistake.

"Nevertheless animals in a free, wild state are perhaps less subject than any other living beings, man not excepted, to alterations, changes, and variations of all kinds. Being free to choose their own food and climate, they vary less than domestic animals vary."[60] The conditions of their existence, in fact, remaining practically constant, the animals are no less constant themselves.

The writer of the above could hardly be claimed as a very thick and thin partisan of immutability, even though he had not shown from the first how clearly he saw that there was no middle position between the denial of all mutability, and the admission that in the course of sufficient time any conceivable amount of mutability is possible. I will give a considerable part of what I have found in the first six volumes of Buffon to bear one way or the other on his views concerning the mutability of species; and I think the reader, so far from agreeing with M. Isidore Geoffroy that Buffon began his work with a belief in the fixity of species, will find, that from the very first chapter onward, he leant strongly to mutability, even if he did not openly avow his belief in it.

In support of this assertion, one quotation must suffice:--

"Nature advances by gradations which pass unnoticed. She passes from one species, and often from one genus to another by imperceptible degrees, so that we meet with a great number of mean species and objects of such doubtful characters that we know not where to place them."[61]

The reader who turns to Buffon himself will find the idea that Buffon took a less advanced position in his old age than he had taken in middle life is also without foundation.

Mr. Darwin has said that Buffon "does not enter into the causes or means of the transformation of species." It is not easy to admit the justice of this. Independently of his frequently insisting on the effect of all kinds of changed surroundings, he has devoted a long chapter of over sixty quarto pages to this very subject; it is to be found in his fourteenth volume, and is headed "De la Dégénération des Animaux," of which words "On descent with modification" will be hardly more than a literal translation. I shall give a fuller but still too brief outline of the chapter later on, and will confine myself here to saying that the three principal causes of modification which Buffon brings forward are changes of climate, of food, and the effects of domestication. He may be said to have attributed variation to the direct and specific action of changed conditions of life, and to have had but little conception of the view which he was himself to suggest to Dr. Erasmus Darwin, and through him to Lamarck.

Isidore Geoffroy, writing of Lamarck, and comparing his position with that taken by Buffon, says, on the whole truly, that "what Buffon ascribes to the general effects of climate, Lamarck maintains to be caused, especially in the case of animals, by the force of habits; *so that, according to him, they are not, properly speaking, modified by the conditions of their existence, but are only induced by these conditions to set about modifying themselves*."[62] But it is very hard to say how much Buffon saw and how much he did not see. He may be trusted to have seen that if he once allowed the thin end of this wedge into his system, he could no more assign limits to the effect which living forms might produce upon their own organisms by effort and ingenuity in the course of long time, than he could set limits to what he had called the power of Nature if he was once to admit that an ass and a horse might, through that power, have been descended from a common ancestor. Nevertheless, he shows no unwillingness or recalcitrancy about letting the wedge enter, for he speaks of domestication as inducing modifications "sufficiently profound to become constant and hereditary in successive generations ... *by its action on bodily habits it influences also their natures, instincts, and most inward qualities*."[63]

This is a very thick thin end to have been allowed to slip in unawares; but it is astonishing how little Buffon can see when he likes. I hardly doubt but he would have been well enough pleased to have let the wedge enter still farther, but this fluctuating writer had assigned himself his limits some years before, and meant adhering to them. Again, in this very chapter on Degeneration, to which M. Geoffroy has referred, there are passages on the callosities on a camel's knees, on the llama, and on the haunches of pouched monkeys which might have been written by Dr. Darwin himself.[64] They will appear more fully presently. Buffon now probably felt that he had said enough, and that others might be trusted to carry the principle farther when the time was riper for its enforcement.

FOOTNOTES:

[51] 'Origin of Species,' p. xiii. ed. 1876.

[52] 'Hist. Nat. Gén.,' tom. ii. p. 405, 1859.

[53] 'Origin of Species,' p. xiv. 1876.

[54] 'Hist. Nat. Gén.,' tom. ii. p. 383.

[55] Tom. iv.

[56] 'Hist. Nat. Gén.,' tom. ii. p. 391, 1859.

[57] Tom. v. p. 59, 1755.

[58] Tom. v. p. 60.

[59] Tom. vi. p. 58, 1756.

[60] Tom. vi. pp. 59-60, 1756.

[61] Tom. i. p. 13, 1749.

[62] 'Hist. Nat. Gén.,' tom. ii. p. 411, 1859.

[63] Tom. xi. p. 290, 1764 (misprinted on title-page 1754).

[64] See tom. xiv. p. 326, 1766; and p. 162 of this volume.

CHAPTER XI.

BUFFON--FULLER QUOTATIONS.

Let us now proceed to those fuller quotations which may answer the double purpose of bearing me out in the view of Buffon's work which I have taken in the foregoing pages, and of inducing the reader to turn to Buffon himself.

I have already said that from the very commencement of his work Buffon showed a proclivity towards considerations which were certain to lead him to a theory of evolution, even though he had not, as I believe he had, already taken a more comprehensive view of the subject than he thought fit to proclaim unreservedly.

In 1749, at the beginning of his first volume he writes:--

"'The first truth that makes itself apparent on serious study of Nature, is one that man may perhaps find humiliating; it is this--that he, too, must take his place in the ranks of animals, being, as he is, an animal in every material point. It is possible also that the instinct of the lower animals will strike him as more unerring, and their industry more marvellous than his own. Then, running his eye over the different objects of which the universe is composed, he will observe with astonishment that we can descend by almost imperceptible degrees from the most perfect creature to the most formless matter--from the most highly organized animal to the most entirely inorganic substance. He will recognize this gradation as the great work of Nature; and he will observe it not only as regards size and form, but also in respect of movements, and in the successive generations of every species.[65]

"Hence," he continues, "arises the difficulty of arriving at any perfect system or method in dealing either with Nature as a whole or even with any single one of her subdivisions. The gradations are so subtle that we are often obliged to make arbitrary divisions. Nature knows nothing about our classifications, and does not choose to lend herself to them without reserve. We therefore see a number of intermediate species and objects which it is very hard to classify, and which of necessity derange our system whatever it may be."[66]

"The attempt to form perfect systems has led to such disastrous results that it is now more easy to learn botany than the terminology which has been adopted as its language."[67]

After saying that "*la marche de la Nature*" has been misunderstood, and that her progress has ever been by a succession of slow steps, he maintains that the only proper course is to class together whatever objects resemble one another, and to separate those which are unlike. If individual specimens are absolutely alike, or differ so little that the differences can hardly be perceived, they must be classed as of the same species; if the differences begin to be perceptible, but if at the same time there is more resemblance than difference, the individuals presenting these features should be classed as of a different species, but as of the same genus; if the differences are still more marked, but nevertheless do not exceed the resemblances, then they must be taken as not only specific but generic, though as not sufficient to warrant the individuals in which they appear, being placed in different classes. If they are still greater, then the individuals are not even of the same class; but it should be always understood that the resemblances and differences are to be considered in reference to the entirety of the plant or animal, and not in reference to any particular part only.[68] The two rocks which are equally to be avoided are, on the one hand, absence of method, and, on the other, a tendency to over-systematize.[69]

Like Dr. Erasmus Darwin, and more recently Mr. Francis Darwin, Buffon is more struck with the resemblances than with the differences between animals and plants, but he supposes the vegetable kingdom to be a continuation of the animal, extending lower down the scale, instead of holding as Dr. Darwin did, that animals and vegetables have been contemporaneous in their degeneration from a common stock.

"We see," he writes, "that there is no absolute and essential difference between animals and vegetables, but that Nature descends by subtle gradations from what we deem the most perfect animal to one which is less so, and again from this to the vegetable. The fresh-water polypus may perhaps be considered as the lowest animal, and as at the same time the highest plant."[70]

Looking to the resemblances between animals and plants, he declares that their modes of reproduction and growth involve such close analogy that no difference of an essential nature can be admitted between them.[71]

On the other hand, Buffon appears, at first sight, to be more struck with the points of difference between the mental powers of the lower animals and man than with those which they present in common. It is impossible, however, to accept this as Buffon's real opinion, on the strength of isolated passages, and in face of a large number of others which point stealthily but irresistibly to an exactly opposite conclusion. We find passages which show a clear apprehension of facts that the world is only now beginning to consider established, followed by others which no man who has kept a dog or cat will be inclined to agree with. I think I have already explained this sufficiently by referring it to the impossibility of his taking any other course under the circumstances of his own position and the times in which he lived. Buffon does not deal with such pregnant facts, as, for example, the geometrical ratio of increase, in such manner as to suggest that he was only half aware of their importance and bearing. On the contrary, in the very middle of those passages which, if taken literally, should most shake confidence in his judgment, there comes a sustaining sentence, so quiet that it shall pass unnoticed by all who are not attentive listeners, yet so encouraging to those who are taking pains to understand their author that their interest is revived at once.

Thus, he has insisted, and means insisting much further, on the many points of resemblance between man and the lower animals, and it has now become necessary to neutralize the effect of what he has written upon the minds of those who are not yet fitted to see instinct and reason as differentiations of a single faculty. He accordingly does this, and, as is his wont, he does it handsomely; so handsomely that even his most admiring followers begin to be uncomfortable. Whereon he begins his next paragraph with "Animals have excellent senses, but not *generally, all of them*, as good as man's."[72] We have heard of damning with faint praise. Is not this to praise with faint damnation? Yet we can lay hold of nothing. It was not Buffon's intention that we should. An ironical writer, concerning whom we cannot at once say whether he is in earnest or not, is an actor who is continually interrupting his performance in order to remind the spectator that he is acting. Complaint, then, against an ironical writer on the score that he puzzles us, is a complaint against irony itself; for a writer is not ironical unless he puzzles. He should not puzzle unless he believes that this is the best manner of making his reader understand him in the end, or without having a *bonne bouche* for those who will be at the pains to puzzle over him; and he should make it plain that for long parts of his work together he is to be taken according to the literal interpretation of his words; but if he has observed the above duly, he is a successful or unsuccessful writer according as he puzzles or fails to do so, and should be praised or blamed accordingly. To condemn irony entirely, is to say that there should be no people allowed to go about the world but those to whom irony would be an impertinence.

Having already in some measure reassured us by the faintness with which he disparages the senses of the lower animals, Buffon continues, that these senses, whether in man or in animals, may be greatly developed by exercise: which we may suppose that a man of even less humour than Buffon must know to be great nonsense, unless it be taken to involve that animals as well as man can reflect and remember; it now, therefore, becomes necessary to reassure the other side, and to maintain that animals cannot reflect, and have no memory. "*Je crois*," he writes, "*qu'on peut démontrer que les animaux n'ont aucune connaissance du passé, aucune idée du temps, et que par conséquent ils n'ont pas la mémoire*."[73]

I am ashamed of even arguing seriously against the supposition that this was Buffon's real opinion. The very sweepingness of the assertion, the baldness, and I might say brutality with which it is made, are convincing in their suggestiveness of one who is laughing very quietly in his sleeve.

"Society," he continues, later on, "considered even in the case of a single human family, involves the power of reason; it involves feeling in such of the lower animals as form themselves into societies freely and of their own accord, but it involves nothing whatever in the case of bees, who have found themselves thrown together through no effort of their own. Such societies can only be, and it is plain have only been, the results--neither foreseen, nor ordained, nor conceived by those who achieve them--of the universal mechanism and of the laws of movement established by the Creator."[74] A hive of bees, in fact, is to be considered as composed of "ten thousand animated automata."[75] Years later he repeats these views with little if any modification.[76] A still more remarkable passage is to be found a little farther on. "If," he asks, "animals have neither understanding, mind, nor memory, if they are wholly without intelligence, and if they are limited to the exercise and experience of feeling only," and it must be remembered that Buffon has denied all these powers to the inferior animals, "whence comes that remarkable prescient instinct which so many of them exhibit? Is the mere power of feeling sensations sufficient to make them garner up food during the summer, on which food they may subsist in winter? Does not this involve the power of comparing dates, and the idea of a coming future, an '*inquiétude raisonnée*'? Why do we find in the hole of the field-mouse enough acorns to keep him until the following summer? Why do we find such an abundant store of honey and wax within the bee-hive? Why do ants store food? Why should birds make nests if they do not know that they will have need of them? Whence arise the stories that we hear of the wisdom of foxes, which hide their prey in different spots, that they may find it at their need and live upon it for days together? Or of the subtilty of owls, which husband their store of mice by biting off their feet, so that they cannot run away? Or of the marvellous penetration of bees, which know beforehand that their queen should lay so many eggs in such and such a time, and that so many of these eggs should be of a kind which will develop into drones, and so many more of such another kind as should become neuters; and who in consequence of this their foreknowledge build so many larger cells for the first, and so many smaller for the second?"[77]

Buffon answers these questions thus:--

"Before replying to them," he says, "we should make sure of the facts themselves;--are they to be depended upon? Have they been narrated by men of intelligence and philosophers, or are they popular fables only?" (How many delightful stories of the same character does he not soon proceed to tell us himself). "I am persuaded that all these pretended wonders will disappear, and the cause of each one of them be found upon due examination. But admitting their truth for a moment, and granting to the narrators of them that animals have a presentiment, a forethought, and even a certainty concerning coming events, does it therefore follow that this should spring from intelligence? If so, theirs is assuredly much greater than our own. For our foreknowledge amounts to conjecture only; the vaunted light of our reason doth but suffice to show us a little probability; whereas the forethought of animals is unerring, and must spring from some principle far higher than any we know of through our own experience. Does not such a consequence, I ask, *prove repugnant alike to religion and common sense*?"[78]

This is Buffon's way. Whenever he has shown us clearly what we ought to think, he stops short suddenly on religious grounds. It is incredible that the writer who at the very commencement of his work makes man take his place among the animals, and who sees a subtle gradation extending over all living beings "from the most perfect creature"--who must be man--"to the most entirely inorganic substance"--I say it is incredible that such a writer should not see that he had made out a stronger case in favour of the reason of animals than against it.

According to him, the test whether a thing is to have such and such a name is whether it looks fairly like other things to which the same name is given; if it does, it is to have the name; if it does not, it is not. No one accepted this lesson more heartily than Dr. Darwin, whose shrewd and homely mind, if not so great as Buffon's, was still one of no common order. Let us see the view he took of this matter. He writes:--

"If we were better acquainted with the histories of those insects which are formed into societies, as the bees, wasps, and ants, I make no doubt but we should find that their arts and improvements are not so similar and uniform as they now appear to us, but that they arose in the same manner from experience and tradition, as the arts of our own species; though their reasoning is from fewer ideas, is busied about fewer objects, and is executed with less energy."[79]

And again, a little later:--

"According to the late observations of Mr. Hunter, it appears that beeswax is not made from the dust of the anthers of flowers, which they bring home on their thighs, but that this makes what is termed bee-bread, and is used for the purpose of feeding the bee-maggots; in the same way butterflies live on honey, but the previous caterpillar lives on vegetable leaves, while the maggots of large flies require flesh for their food. What induces the bee, who lives on honey, to lay up vegetable powder for its young? What induces the butterfly to lay its eggs on leaves when itself feeds on honey?... If these are not deductions from their own previous experience or observation, all the actions of mankind must be resolved into instincts."[80]

Or again:--

"Common worms stop up their holes with leaves or straws to prevent the frost from injuring them, or the centipes from devouring them. The habits of peace or the stratagems of war of these subterranean nations are covered from our view; but a friend of mine prevailed on a distressed worm to enter the hole of another worm on a bowling green, and he presently returned much wounded about the head, ... which evinces they have design in stopping the mouths of their habitations."[81]

Does it not look as if Dr. Darwin had in his mind the very passage of Buffon which I have been last quoting? and is it likely that the facts which were accepted by Dr. Darwin without question, or the conclusions which were obvious to him, were any less accepted by or obvious to Buffon?

The Goat--Hybridism.

In his prefatory remarks upon the goat, Buffon complains of the want of systematic and certified experiment as to what breeds and species will be fertile *inter se*, and with what results. The passage is too long to quote, but is exceedingly good, and throughout involves belief in a very considerable amount of modification in the course of successive generations. I may give the following as an example:--

"We do not know whether or no the zebra would breed with the horse or ass--whether the large-tailed Barbary sheep would be fertile if crossed with our own--whether the chamois is not a wild goat; and whether it would not form an intermediate breed if crossed with our domesticated goats; we do not know whether the differences between apes are really specific, or whether apes are not like dogs, one single species, of which there are many different breeds.... Our ignorance concerning all these facts is almost inevitable, as the experiments which would decide them require more time, pains, and money than can be spared from, the life and fortune of an ordinary man. I have spent many years in experiments of this kind, and will give my results when I come to my chapter on mules; but I may as well say at once that they have thrown but little light upon the subject, and have been for the most part unsuccessful."[82]

"But these," he continues, "are the very points which must determine our whole knowledge concerning animals, their right division into species, and the true understanding of their history." He proposes therefore, in the present lack of knowledge, "to regard all animals as different species which do not breed together under our eyes," and to leave time and experiment to correct mistakes.[83]

The Pig--Doctrine of Final Causes.

We have seen that the doctrine of the mutability of species has been unfortunately entangled with that of final causes, or the belief that every organ and every part of each animal or plant has been designed to serve some purpose useful to the animal, and this not only useful at some past time, but useful now, and for all time to come. He who believes species to be mutable will see in many organs signs of the history of the individual, but nothing more. Buffon, as I have said, is explicit in his denial of final causes in the sense expressed above. After pointing out that the pig is an animal whose relation to other animals it is difficult to define, he says:--

"In a word, it is of a nature altogether equivocal and ambiguous, or, rather, it must appear so to those who believe the hypothetical order of their own ideas to be the real order of things, and who see nothing in the infinite chain of existences but a few apparent points to which they will refer everything.

"But we cannot know Nature by inclosing her action within the narrow circle of our own thoughts.... Instead of limiting her action, we should extend it through immensity itself; we should regard nothing as impossible, but should expect to find all things--supposing that all things are possible--nay, *are*. Doubtful species, then, irregular productions, anomalous existences will henceforth no longer surprise us, and will find their place in the infinite order of things as duly as any others. They fill up the links of the chain; they form knots and intermediate points, and also they mark its extremities: they are of especial value to human intelligence, as providing it with cases in which Nature, being less in conformity with herself, is taken more unawares, so that we can recognize singular characters and fleeting traits which show us that her ends are much more general than are our own views of those ends, and that, though she does nothing in vain, yet she does but little with the designs which we ascribe to her."[84]

"The pig," he continues, "is not formed on an original, special, and perfect type; its type is compounded of that of many other animals. It has parts which are evidently useless, or which at any rate it cannot use--such as toes, all the bones of which are perfectly formed but which are yet of no service to it. Nature then is far from subjecting herself to final causes in the composition of her creatures. Why should she not sometimes add superabundant parts, seeing she so often omits essential ones?" "How many animals are there not which lack sense and limbs? Why is it considered so necessary that every part in an individual should be useful to the other parts and to the whole animal? Should it not be enough that they do not injure each other nor stand in the way of each other's fair development? All parts coexist which do not injure each other enough to destroy each other, and perhaps in the greater number of living beings the parts which must be considered as relative, useful, or necessary, are fewer than those which are indifferent, useless, and superabundant. But we--ever on the look out to refer all parts to a certain end--when we can see no apparent use for them suppose them to have hidden uses, and imagine connections which are without foundation, and serve only to obscure our perception of Nature as she really is: we fail to see that we thus rob philosophy of her true character, which is to inquire into the 'how' of things--into the manner in which Nature acts--and that we substitute for this true object a vain idea, seeking to divine the 'why'--the ends which she has proposed in acting."[85]

The Dog--Varieties in consequence of Man's Selection.

"Of all animals the dog is most susceptible of impressions, and becomes most easily modified by moral causes. He is also the one whose nature is most subject to the variations and alterations caused by physical influences: he varies to a prodigious extent, in temperament, mental powers, and in habits: his very form is not constant;" ... but presents so many differences that "dogs have nothing in common but conformity of interior organization, and the power of interbreeding freely."...

... "How then can we detect the characters of the original race? How recognize the effects produced by climate, food, &c.? How, again, distinguish these from those other effects which come from the intermixture of races, either when wild or in a state of domestication? All these causes, in the course of time, alter even the most constant forms, so that the imprint of Nature does not preserve its sharpness in races which man has dealt with largely. Those animals which are free to choose climate and food for themselves can best conserve their original character, ... but those which man has subjected to his own influence--which he has taken with him from clime to clime, whose food, habits, and manner of life he has altered--must also have changed their form far more than others; and as a matter of fact we find much greater variety in the species of domesticated animals than in those of wild ones. Of all these, however, the dog is the one most closely attached to man, living like man the least regular manner of life; he is also the one whose feelings so master him as to make him docile, obedient, susceptible of every kind of impression, and even of every kind of constraint; it is not surprising, then, that he should of all animals present us with the greatest variety in shape, stature, colour, and all physical and mental qualities."

Here again the direct cause of modification is given as being the inner feelings of the animal modified, change of conditions being the indirect cause as with Dr. Erasmus Darwin and Lamarck.

"Other circumstances, however, concur to produce these results. The dog is short-lived: he breeds often and freely: he is perpetually under the eye of man; hence when--by some chance common enough with Nature--a variation or special feature has made its appearance, man has tried to perpetuate it by uniting together the individuals in which it has appeared, as people do now who wish to form new breeds of dogs and other animals. Moreover, though species were all formed at the same time, yet the number of generations since the creation has been much greater in the short-lived than in the long-lived species: hence variations, alterations, and departure from the original type, may be expected to have become more perceptible in the case of animals which are so much farther removed from their original stock.

"Man is now eight times nearer Adam than the dog is to the first dog--for man lives eighty years, while the dog lives but ten. If, then, these species have an equal tendency to depart from their original type, the departure should be eight times more apparent with the dog than with man."[86]

Here follow remarks upon the great variability of ephemeral insects and of animal plants, on the impossibility of discovering the parent-stock of our wheat and of others of our domesticated plants,[87] and on the tendency of both plants and animals to resume feral characteristics on becoming wild again after domestication.[88]

The Hare--Geometrical Ratio of Increase.

We have already seen that it was Buffon's pleasure to consider the hare a rabbit for the time being, and to make it the text for a discourse upon fecundity. I have no doubt he enjoyed doing this, and would have found comparatively little pleasure in preaching the same discourse upon the rabbit. Speaking of the way in which even the races of mankind have struggled and crowded each other out, Buffon says:--

"These great events--these well-marked epochs in the history of the human race--are yet but ripples, as it were, on the current of life; which, as a general rule, flows onward evenly and in equal volume.

"It may be said that the movement of Nature turns upon two immovable pivots--one, the illimitable fecundity which she has given to all species; the other, the innumerable difficulties which reduce the results of that fecundity, and leave throughout time nearly the same quantity of individuals in every species.[89]... Taking the earth as a whole, and the human race in its entirety, the numbers of mankind, like those of animals, should remain nearly constant throughout time; for they depend upon an equilibrium of physical causes which has long since been reached, and which neither man's moral nor his physical efforts can disturb, inasmuch as these moral efforts do but spring from physical causes, of which they are the special effects. No matter what care man may take of his own species, he can only make it more abundant in one place by destroying it or diminishing its numbers in another. When one part of the globe is overpeopled, men emigrate, spread themselves over other countries, destroy one another, and establish laws and customs which sometimes only too surely prevent excess of population. In those climates where fecundity is greatest, as in China, Egypt, and Guinea, they banish, mutilate, sell, or drown infants. Here, we condemn them to a perpetual celibacy. Those who are in being find it easy to assert rights over the unborn. Regarding themselves as the necessary, they annihilate the contingent, and suppress future generations for their own pleasure and advantage. Man does for his own race, without perceiving it, what he does also for the inferior animals: that is to say, he protects it and encourages it to increase, or neglects it according to his sense of need--according as advantage or inconvenience is expected as the consequence of either course. And since all these moral effects themselves depend upon physical causes, which have been in permanent equilibrium ever since the world was formed, it follows that the numbers of mankind, like those of animals, should remain constant.

"Nevertheless, this fixed state, this constant number, is not absolute, all physical and moral causes, and all the results which spring from them, balance themselves, as though, upon a see-saw, which has a certain play, but never so much as that equilibrium should be altogether lost. As everything in the universe is in movement, and as all the forces which are contained in matter act one against the other and counterbalance one another, all is done by a kind of oscillation; of which the mean points are those to which we refer as being the ordinary course of nature, while the extremes are the periods which deviate from that course most widely. And, as a matter of fact, with animals as much as with plants, a time of unusual fecundity is commonly followed by one of sterility; abundance and dearth come alternately, and often at such short intervals that we may foretell the production of a coming year by our knowledge of the past one. Our apples, pears, oaks, beeches, and the greater number of our fruit and forest trees, bear freely but about one year in two. Caterpillars, cockchafers, woodlice, which in one year may multiply with great abundance, will appear but sparsely in the next. What indeed would become of all the good things of the earth, what would become of the useful animals, and indeed of man himself, if each individual in these years of excess was to leave its quotum of offspring? This, however, does not happen, for destruction and sterility follow closely upon excessive fecundity, and, independently of the contagion which follows inevitably upon overcrowding, each species has its own special sources of death and destruction, which are of themselves sufficient to compensate for excess in any past generation.

"Nevertheless the foregoing should not be taken in an absolute sense, nor yet too strictly,--especially in the case of those races which are not left entirely to the care of Nature. Those which man takes care of--commencing with his own--are more abundant than they would be without his care, yet, as his power of taking this care is limited, the increase which has taken place is also fixed, and has long been restrained within impassable boundaries. Again, though in civilized countries man, and all the animals useful to him, are more numerous than in other places, yet their numbers never become excessive, for the same power which brings them into being destroys them as soon as they are found inconvenient."[90]

The Carnivora--Sensation.

Buffon begins his seventh volume with some remarks on the *carnivora* in general, which I would gladly quote at fuller length than my space will allow. He dwells on the fact that the number, as well as the fecundity of the insect races is greater than that of the mammalia, and even than of plants; and he points out that "violent death is almost as necessary an usage as is the law that we must all, in one way or another, die." This leads him to the question whether animals can feel. "To speak seriously," (au réel) he says (and why this, if he had always spoken seriously?[91]), "can we doubt that those animals whose organization resembles our own, feel the same sensations as we do? They must feel, for they have senses, and they must feel more and more in proportion as their senses are more active and more perfect." Those whose organ of any sense is imperfect, have but imperfect perception in respect of that sense; and those that are entirely without the organ want also all corresponding sensation. "Movement is the necessary consequence of acts of perception. I have already shown that in whatever manner a living being is organized, if it has perceptions at all, it cannot fail to show that it has them by some kind of movement of its body. Hence plants, though highly organized, have no feeling, any more than have those animals which, like plants, manifest no power of motion. Among animals there are those which, like the sensitive plant, have but a certain power of movement about their own parts, and which have no power of locomotion; such animals have as yet but little perception. Those, again, which have power of locomotion, but which, like automata, do but a small number of things, and always after the same fashion, can have only small powers of perception, and these limited to a small number of objects. But in the case of man, what automata, indeed, have we not here! How much do not education and the intercommunication of ideas increase our powers and vivacity of perception. What difference can we not see in this respect between civilized and uncivilized races, between the peasant girl, and the woman of the world? And in like manner among animals, those which live with us have their perceptions increased in range, while those that are wild have but their natural instinct, which is often more certain but always more limited in range than is the intelligence of domesticated animals."[92]

"For perception to exist in its fullest development in any animal body, that body must form a whole--an *ensemble*, which shall not only be capable of feeling in all its parts, but shall be so arranged that all these feeling parts shall have a close correspondence with one another, and that no one of them can be disturbed without communicating a portion of that disturbance to every other part. There must also be a single chief centre, with which all these different disturbances may be connected, and from which, as from a common *point d'appui*, the reactions against them may take their rise. Hence man, and those animals whose organization most resembles man's, will be the most capable of perceptions, while those whose unity is less complete, whose parts have a less close correspondence with each other--which have several centres of sensation, and which seem, in consequence, less to envelope a single existence in a single body than to contain many centres of existence separated and different from one another--these will have fewer and duller perceptions. The polypus, which can be reproduced by fission; the wasp, whose head even after separation from the body still moves, lives, acts, and even eats as heretofore; the lizard which we deprive neither of sensation nor movement by cutting off part of its body; the lobster which can restore its amputated limbs; the turtle whose heart beats long after it has been plucked out, in a word all the animals whose organization differs from our own, have but small powers of perception, and the smaller the more they differ from us."[93]

This is Buffon's way of satirizing our inability to bear in mind that we are compelled to judge all things by our own standards. He also wishes to reassure those who might be alarmed at the tendency of some of his foregoing remarks, and who he knew would find comfort in being told that a thing which does not express itself as they do does not feel at all.

The diaphragm according to Buffon appears to be the centre of the powers of sensation; the slightest injury "even to the attachments of the diaphragm is followed by strong convulsions, and even by death. The brain which has been called the seat of 'sensations' is yet not the centre of 'perception,' since we can wound it, and even take considerable parts of it away, without death's ensuing, and without preventing an animal from living, moving and feeling in all its parts."

Buffon thus distinguishes between "sensation" and "perception." "Sensation," he says, "is simply the activity of a sense, but perception is the pleasantness or unpleasantness of this sensation," "perceived by its being propagated and becoming active throughout the entire system." I have therefore several times, when translating from Buffon, rendered the word "*sentiment*" by "perception," and shall continue to do so. "I say," writes Buffon, "the pleasantness or unpleasantness, because this is the very essence of perception; the one feature of perception consists in perceiving either pain or pleasure; and though movements which do not affect us in either one or the other of these two ways may indeed take place within us, yet we are indifferent to them, and do not perceive that we are affected by them. All external movement, and all exercise of the animal powers, spring from perception; its action is proportionate to the extent of its excitation, to the extent of the feeling which is being felt.[94] And this same part, which we regard as the centre of sensation, will also be that of all the animal powers; or, if it is preferred to call it so, it will be the common *point d'appui* from which they all take rise. The diaphragm is to the animal what the 'stock' is to the plant; both divide an organism transversely, both serve as the *point d'appui* of opposing forces; for the forces which push upward those parts of a tree which should form its trunk and branches, bear upon and are supported by the 'stock,' as do those opposing forces, which drive the roots downwards.

"Even on a cursory examination we can see that all our innermost affections, our most lively emotions, our most expansive moments of delight, and, on the other hand, our sudden starts, pains, sicknesses, and swoons--in fact, all our strong impressions concerning the pleasure or pain of any sensation--make themselves felt within the body, and about the region of the diaphragm. The brain, on the contrary, shows no sign of being a seat of perception. In the head there are pure sensations and nothing else, or rather, there are but the representations of sensations stripped of the character of perception; that is to say, we can remember and call to mind whether such and such a sensation was pleasant to us or otherwise, and if this operation, which goes on in the head, is followed by a vivid perception, then the impression made is perceived in the interior of the body, and always in the region of the diaphragm. Hence, in the foetus where this membrane is without use, there is no perception, or so little that nothing comes of it, the movements of the foetus, such as they are, being rather mechanical than dependent on sensation and will.

"Whatever the matter may be which serves as the vehicle of perception, and produces muscular movement, it is certain that it is propagated through the nerves, and that it communicates itself instantaneously from one extremity of the system to the other. In whatever manner this operation is conducted, whether by the vibrations, as it were, of elastic cords or by a subtle fire, or by a matter resembling electricity, which not only resides in animal as in all other bodies, but is being continually renewed in them by the movements of the heart and lungs, by the friction of the blood within the arteries, and also by the action of exterior causes upon our organs of sense--in whatever manner, I say, the operation is conducted, it is nevertheless certain that the nerves and membranes are the only parts in an animal body that can feel. The blood, lymphs, and all other fluids, the fats, bone, flesh, and all other solids, are of themselves void of sensation. And so also is the brain; it is a soft and inelastic substance, incapable therefore of producing or of propagating the movement, vibrations, or concussions which, result in perception. The meninges, on the other hand, are exceedingly sensitive, and are the envelopes of all the nerves; like the nerves, they take rise in the head; and, dividing themselves like the branches of the nerves, they extend even to their smallest ramifications: they are, so to speak, flattened nerves; they are of the same substance as the nerves, are nearly of the same degree of elasticity, and form a necessary part of the system of sensation. If, then, the seat of the sensations must be placed in the head, let it be placed in the meninges, and not in the medullary part of the brain, which is of an entirely different substance."[95]

If this is so, it appears from what will follow as though the meninges must be the "stock" rather than the diaphragm.

"What perhaps has given rise to the opinion that the seat of all sensations and the centre of all sensibility is in the brain, is the fact that the nerves, which are the organs of perception, all attach themselves to the brain, which has hence come to be regarded as the one common centre which can receive all their vibrations and impressions. This fact alone has sufficed to indicate the brain as the origin of perceptions--as the essential organ of sensations; in a word, as the common sensorium. This supposition has appeared so simple and natural that its physical impossibility has been overlooked, an impossibility, however, which should be sufficiently apparent. For how can a part which cannot feel--a soft inactive substance like the brain--be the very organ of perception and movement? How can this soft and perceptionless part not only receive impressions, but preserve them for a length of time, and transmit their undulatory movements (*en propage les ébranlements*) throughout all the solid and feeling parts of the body? It may perhaps be maintained with Descartes and M. de Peyronie that the principle of sensation does not reside in the brain, but in the pineal gland or in the *corpus callosum*; but a glance at the conformation of the brain itself will suffice to show that these parts do not join on to the nerves, but that they are entirely surrounded by those parts of the brain which do not feel, and are so separated from the nerves that they cannot receive any movement from them; whence it follows that this second supposition is as groundless as the first."[96]

What, then, asks Buffon, *is* the use of the brain? Man, the quadrupeds, and birds all have larger brains, and at the same time more extended perceptions, than fishes, insects, and those other living beings whose brains are smaller in proportion. "When the brain is compressed, there is suspension of all power of movement. If this part is not the source of our powers of motion, why is it so necessary and so essential? Why, again, does it seem so proportionate in each animal to the amount of perceiving power which that animal possesses?

"I think I can answer this question in a satisfactory manner, difficult though it seems; but in order that I may do so, I would ask the reader to lend me his attention for a few moments while we regard the brain simply *as brain*, and have no other idea concerning it than we can derive from inspection and reflection. The brain, as well as the *medulla oblongata* and the spinal marrow, which are but prolongations of the brain itself, is only a kind of hardly organized mucilage; we find in it nothing but the extremities of small arteries, which run into it in very great numbers, but which convey a white and nourishing lymph instead of blood. When the parts of the brain are disunited by maceration, these same small arteries, or lymphatic vessels, appear as very delicate threads throughout their whole length. The nerves, on the contrary, do not penetrate the substance of the brain; they abut upon its surface only; before reaching it they lose their elasticity and solidity, and the extremities of the nerves which are nearest to the brain are soft, and nearly mucilaginous. From this exposition, in which there is nothing hypothetical, it appears that the brain, which is nourished by the lymphatic arteries, does in its turn provide nourishment for the nerves, and that we must regard these as a kind of vegetation which rises as trunks and branches from the brain, and become subsequently subdivided into an infinite number, as it were, of twigs. The brain is to the nerves what the earth is to plants: the last extremities of the nerves are the roots, which with every vegetable are more soft and tender than the trunk or branches; they contain a ductile matter fit for the growth and nourishment of the nervous tree or fibre; they draw the ductile matter from the substance of the brain itself, to which the arteries are continually bringing the lymph that is necessary to supply it. The brain, then, instead of being the seat of the sensations, and the originator of perception, is an organ of secretion and nutrition only, though a very essential organ, without which the nerves could neither grow nor be maintained.

"This organ is greater in man, in quadrupeds, and in birds, because the number or bulk of the nerves is greater in these animals than in fishes or insects, whose power of perception is more feeble, for this very reason, that they have but a small brain; one, in fact, that is proportioned to the small quantity of nerves which that brain must support. Nor can I omit to state here that man has not, as has been pretended by some, a larger brain than has any other animal; for there are apes and cetacea which have more brain than man in proportion to the volume of their bodies--another fact which proves that the brain is neither the seat of sensations nor the originator of perception, since in that case these animals would have more sensations and perception than man.

"If we consider the manner in which plants derive their nourishment, we shall find that they do not draw up the grosser parts either of earth or water; these parts must be reduced by warmth into subtle vapours before the roots can suck them up into the plant. In like manner the nutrition of the nerves is only effected by means of the more subtle parts of the humidity of the brain, which are sucked up by the roots or extremities of the nerves, and are carried thence through all the branches of the sensory system. This system forms, as we have said, a whole, all whose parts are interconnected by so close a union that we cannot wound one without communicating a violent shock to all the others; the wounding or simply pulling of the smallest nerve is sufficient to cause lively irritation to all the others, and to put the body in convulsion; nor can we ease this pain and convulsion except by cutting the nerve higher up than the injured part; but on this all the parts abutting on this nerve become thenceforward senseless and immovable for ever. The brain should not be considered as of the same character, nor as an organic portion of the nervous system, for it has not the same properties nor the same substance, being neither solid nor elastic, nor yet capable of feeling. I admit that on its compression perception ceases, but this very fact shows it to be a body foreign to the nervous system itself, which, acting by its weight, or pressure, against the extremities of the nerves, oppresses them and stupefies them in the same way as a weight placed upon the arm, leg, or any other part of the body, stupefies the nerves and deadens the perceptions of that part. And it is evident that this cessation of sensation on compression is but a suspension and temporary stupefaction, for the moment the compression of the brain ceases, perception and the power of movement returns. Again, I admit that on tearing the medullary substance, and on wounding the brain till the *corpus callosum* is reached, convulsion, loss of sensation, and death ensue; but this is because the nerves are so entirely deranged that they are, so to speak, torn up by the roots and wounded all together, and at their source.

"In further proof that the brain is neither the centre of perception nor the seat of the sensations, I may remind the reader that animals and even children have been born without heads and brains, and have yet had feeling, movement, and life. There are also whole classes of animals, like insects and worms, with a brain that is by no means a distinct mass nor of sensible volume, but with only something which corresponds with the *medulla oblongata* and the spinal marrow. There would be more reason, then, in placing the seat of the feelings and perceptions in the spinal marrow, which no animal is without, than in the brain which is not an organ common to all creatures that can feel."

If Buffon's ideas concerning the brain are as just as they appear to be, the resemblance between plants and animals is more close than is apparent, even to a superficial observer, on a first inspection of the phenomena. Such an observer, however, on looking but a little more intently, will see the higher *vertebrata* as perambulating vegetables planted upside down. So the man who had been born blind, on being made to see, and on looking at the objects before him with unsophisticated eyes, said without hesitation that he saw "men as trees walking," thus seeing with more prophetic insight than either he or the bystanders could interpret. For our skull is as a kind of flower-pot, and holds the soil from which we spring, that is to say the brain; our mouth and stomach are roots, in two stories or stages; our bones are the trellis-work to which we cling while going about in search of sustenance for our roots; or they are as the woody trunk of a tree; *we* are the nerves which are rooted in the brain, and which draw thence the sustenance which is supplied it by the stomach; our lungs are leaves which are folded up within us, as the blossom of a fig is hidden within the fruit itself.

This is what should follow if Buffon's theory of the brain is allowed to stand, which I hope will prove to be the case, for it is the only comfortable thought concerning the brain that I have met with in any writer. I have given it here at some length on account of its importance, and for the illustration it affords of Buffon's hatred of mystery, rather than for its bearing upon evolution. The fact that our leading men of science have adopted other theories will weigh little with those who have watched scientific orthodoxy with any closeness. What Buffon thought of that orthodoxy may be gathered from the following:--

"The greatest obstacles to the advancement of human knowledge lie less in things themselves than in man's manner of considering them. However complicated a machine the human body may be, it is still less complicated than are our own ideas concerning it. It is less difficult to see Nature as she is, than as she is presented to us. She carries a veil only, while we would put a mask over her face; we load her with our own prejudices, and suppose her to act and to conduct her operations even after the same fashion as ourselves.[97]

"I am by no means speaking of those purely arbitrary systems which we are able at a glance to detect as chimeras that are being pretended to us as realities, but I refer to the methods whereby people have set themselves seriously to study nature. Even the experimental method itself has been more fertile of error than of truth, for though it is indeed the surest, yet is it no surer than the hand of him who uses it. No matter how little we incline out of the straight path, we soon find ourselves wandering in a sterile wilderness, where we can see but a few obscure objects scattered sparsely; nevertheless we do violence to these facts and to ourselves, and resemble them together on a conceit of analogies and common properties amongst them. Then, passing and repassing complaisantly over the tortuous path which we have ourselves beaten, we deem the road a worn one, and though it leads no whither, the world follows it, adopts it, and accepts its supposed consequences as first principles. I could show this by laying bare the origin of that which goes by the name of 'principle' in all the sciences, whether abstract or natural. In the case of the former, the basis of principle is abstraction--that is to say, one or more suppositions: in that of the second, principles are but the consequences, better or worse, of the methods which may have been followed. And to speak here of anatomy only, did not he who first surmounted his natural repugnance and set himself to work to open a human body--did he not believe that through going all over it, dissecting it, dividing it into all its parts, he would soon learn its structure, mechanism, and functions? But he found the task greater than he had expected, and renouncing such pretensions, was fain to content himself with a method--not for seeing and judging, but for seeing after an orderly fashion. This method ... is still the sole business of our ablest anatomists, but it is not science. It is the road which should lead scienceward, and might perhaps have reached science itself, if instead of walking ever on a single narrow path men had set the anatomy of man and that of animals face to face with one another. For, what real knowledge can be drawn from an isolated pursuit? Is not the foundation of all science seen to consist in the comparison which the human mind can draw between different objects in the matter of their resemblances and differences--of their analogous or conflicting properties, and of all the relations in which they stand to one another? The absolute, if it exist at all, is but of the concurrence of man's own knowledge; we judge and can judge of things only by their bearings one upon another; hence whenever a method limits us to only a single subject, whenever we consider it in its solitude and without regard to its resemblances or to its differences from other objects, we can attain to no real knowledge, nor yet, much less, reach any general principle. We do but give names, and make descriptions of a thing, and of all its parts. Hence comes it that, after three thousand years of dissection, anatomy is still but a nomenclature, and has hardly advanced a step towards its true object, which is the science of animal economy. Furthermore, what defects are there not in the method itself, which should above all things else be simple and easy to be understood, depending as it does upon inspection and having denominations only for its end! For seeing that nomenclature has been mistaken for knowledge, men have made it their chief business to multiply names, instead of limiting things; they have crushed themselves under the burden of details, and been on the look out for differences where there was no distinction. When they had given a new name they conceived of it as a new thing, and described the smallest parts with the most minutious exactness, while the description of some still smaller part, forgotten or neglected by previous anatomists, has been straightway hailed as a discovery. The denominations themselves being often taken from things which had no relation to the object that it was desired to denominate, have served but to confound confusion. The part of the brain, for example, which is called testes and nates, wherein does it so differ from the rest of the brain that it should deserve a name? These names, taken at haphazard or springing from some preconceived opinion, have themselves become the parents of new prejudices and speculations; other names given to parts which have been ill observed, or which are even non-existent, have been sources of new errors. What functions and uses has it not been attempted to foist upon the pineal gland, and on the alleged empty space in the brain which is called the arch, the first of which is but a gland, while the very existence of the other is doubtful,--the empty space being perhaps produced by the hand of the anatomist and the method of dissection."[98]

The Genus felis.

In his preliminary remarks upon the lion, Buffon while still professing to believe in some considerable mutability of species, seems very far from admitting that all living forms are capable of modification. But he has shown us long since how clearly he saw the impossibility of limiting mutability, if he once admitted so much of the thin end of the wedge as that a horse and an ass might be related. It is plain, therefore, that he is not speaking "*au réel*" here, and we accordingly find him talking clap-trap about the nobleness of the lion in having no species immediately allied to it. A few lines lower on he reminds us in a casual way that the ass and horse are related.

He writes:--

"Added to all these noble individual features the lion has also what may be called a *specific* nobility. For I call those species noble which are constant, invariable, and which are above suspicion of having degenerated. These species are commonly isolated, and the only ones of their genus. They are distinguished by such well-marked features that they cannot be mistaken, nor confounded with any other species. To begin for example with man, the noblest of created beings; he is but of a single species, inasmuch as men and women will breed freely *inter se* in spite of all existing differences of race, climate and colour; and also inasmuch as there is no other animal which can claim either a distant or near relationship with him. The horse, on the other hand, is more noble as an individual than as a species, for he has the ass as his near neighbour, *and seems himself to be nearly enough related to it*; ... the dog is perhaps of even less noble species, approaching as he does to the wolf, fox, and jackal, *which we can only consider to be the degenerated species of a single family*"[99]--all which may seem very natural opinions for a French aristocrat in the days before the Revolution, but which cannot for a moment be believed to have been Buffon's own. I have not ascertained the date of Buffon's little quarrel with the Sorbonne, but I cannot doubt that if we knew the inner history of the work we are considering, we should find this passage and others like it explained by the necessity of quieting orthodox adversaries. He concludes the paragraph from which I have just been quoting by saying, "To class man and the ape together, or the lion with the cat, and to say that the lion is a *cat with a mane and a long tail*--this were to degrade and disfigure nature instead of describing her and denominating her species." Buffon very rarely uses italics, but those last given are his, not mine; could words be better chosen to make us see the lion and the cat as members of the same genus? No wonder the Sorbonne considered him an infelicitous writer; why could he not have said "cat," and have done with it, instead of giving a couple of sly but telling touches, which make the cat as like a lion as possible, and then telling us that we must not call her one? Sorbonnes never do like people who write in this way.

"The lion, then, belongs to a most noble species, standing as he does alone, and incapable of being confounded with the tiger, leopard, ounce, &c., while, on the contrary, those species, which appear to be least distant from the lion, are very sufficiently indistinguishable, so that travellers and nomenclators are continually confounding them."[100]

If this is not pure malice, never was a writer more persistently unfortunate in little ways. Why remind us here that the species which come nearest to the lion are so hard to distinguish? Why not have said nothing about it? As it is, the case stands thus: we are required to admit close resemblance between the leopard and the tiger, while we are to deny it between the tiger and the lion, in spite of there being no greater outward difference between the first than between the second pair, and in spite of the hurried whisper "*cat with a mane and a long tail*" still haunting our ears. Isidore Geoffroy and his followers may consent to this arrangement, but I hope the majority of my readers will not do so.

I went on to the account of the tiger with some interest to see the line which Buffon would take concerning it. I anticipated that we should find cats, pumas, lynxes, &c., to be really very like tigers, and was surprised to learn that the "true" tiger, though certainly not unlike these animals, was still to be distinguished from "many others which had since been called tigers." He is on no account to be confounded with these, in spite of the obvious temptation to confound him. He is "a rare animal, little known to the ancients, and badly described by the moderns." He is a beast "of great ferocity, of terrible swiftness, and surpassing even the proportions of the lion." The effect of the description is that we no longer find the lion standing alone, but with the tiger on a par with him if not above him; but at the same time we fall easy victims to the temptation to confound the tiger with "the many other animals which are also called tigers." A surface stream has swept the members of the cat family in different directions, but a stealthy undercurrent has seized them from beneath, and they are now happily reunited.

Animals of the Old and New World--Changed Geographical Distribution.

Writing upon the animals of the old world,[101] and referring to the humps of the camel and the bison, Buffon shows that very considerable modification may be effected in some animals within even a few generations, but he attributes the effect produced to the direct influence of climate. Buffon concludes his sketch of the animals of the new world by pointing out that the larger animals of the African torrid zone have been hindered by sea and desert from finding their way to America, and by claiming to be the first "even to have suspected" that there was not a single denizen of the torrid zone of one continent which was common also to the other.[102]

The animals common to both continents are those which can stand the cold and which are generally suited for a temperate climate. These, Buffon believes, to have travelled either over some land still unknown, or "more probably," over territory which has long since been submerged. The species of the old and new world are never without some well-marked difference, which however should not be held sufficient for us to refuse to admit their practical identity. But he maintains, I imagine wilfully, that there is a tendency in all the mammalia to become smaller on being transported to the new world, and refers the fact to the quality of the earth, the condition of the climate, the degrees of heat and humidity, to the height of mountains, amounts of running or stagnant waters, extent of forest, and above all to the brutal condition of nature in a new country, which he evidently regards with true aristocratic abhorrence.[103]

Then follows a passage which I had better perhaps give in full:--

The mammoth "was certainly the greatest and strongest of all quadrupeds; but it has disappeared; and if so, how many smaller, feebler, and less remarkable species must have also perished without leaving us any traces or even hints of their having existed? How many other species have changed their nature, that is to say, become perfected or degraded, through great changes in the distribution of land and ocean, through the cultivation or neglect of the country which they inhabit, through the long-continued effects of climatic changes, so that they are no longer the same animals that they once were? Yet of all living beings after man, the quadrupeds are the ones whose nature is most fixed and form most constant: birds and fishes vary much more easily; insects still more again than these, and if we descend to plants, which certainly cannot be excluded from animated nature, we shall be surprised at the readiness with which species are seen to vary, and at the ease with which they change their forms and adopt new natures.

"It is probable then that all the animals of the new world are derived from congeners in the old, without any deviation from the ordinary course of nature. We may believe that having become separated in the lapse of ages, by vast oceans and countries which they could not traverse, they have gradually been affected by, and derived impressions from, a climate which has itself been modified so as to become a new one through the operation of those same causes which dissociated the individuals of the old and new world from one another; thus in the course of time they have grown smaller and changed their characters. This, however, should not prevent our classifying them as different species now, for the difference is no less real whether it is caused by time, climate and soil, or whether it dates from the creation. *Nature I maintain is in a state of continual flux and movement. It is enough for man if he can grasp her as she is in his own time, and throw but a glance or two upon the past and future, so as to try and perceive what she may have been in former times and what one day she may attain to.*"[104]

The Buffalo--Animals under Domestication.

"The bison and the aurochs," says Buffon, "differ only in unessential characteristics, and are, by consequence, of the same species as our domestic cattle, so that I believe all the pretended species of the ox, whether ancient or modern, may be reduced to three--the bull, the buffalo, and the bubalus.

"The case of animals under domestication is in many respects different from that of wild ones; they vary much more in disposition, size and shape, especially as regards the exterior parts of their bodies: the effects of climate, so powerful throughout nature, act with far greater effect upon captive animals than upon wild ones. Food prepared by man, and often ill chosen, combined with the inclemency of an uncongenial climate--these eventuate in modifications sufficiently profound to become constant and hereditary in successive generations. I do not pretend to say that this general cause of modification is so powerful as to change radically the nature of beings which have had their impress stamped upon them in that surest of moulds--heredity; but it nevertheless changes them in not a few respects; it masks and transforms their outward appearance; it suppresses some of their parts, and gives them new ones; it paints them with various colours, and *by its action on bodily habits influences also their natures, instincts, and most inward qualities*" (and what is this but "radically changing their nature"?). "The modification of but a single part, moreover, in a whole as perfect as an animal body, will necessitate a correlative modification in every other part, and it is from this cause that our domestic animals differ almost as much in nature and instinct, as in form, from those from which they originally sprung."[105]

Buffon confirms this last assertion by quoting the sheep as an example--an animal which can now no longer exist in a wild state. Then returning to cattle, he repeats that many varieties have been formed by the effects--"diverse in themselves, and diverse in their combinations--of climate, food, and treatment, whether under domestication or in their wild state." These are the main causes of variation ("causes générales de variété"),[106] among our domesticated animals, but by far the greatest is changed climate in consequence of their accompanying man in his migrations. The effects of the foregoing causes of modification, especially the last of them, are repeatedly insisted on in the course of the forty pages which complete the preliminary account of the buffalo.

What holds good for the buffalo does so also for the mouflon or wild sheep. This, Buffon declares to be the source of all our domesticated breeds: of these there are in all some four or five, "all of them being but degenerations from a single stock, produced by man's agency, and propagated for his convenience."[107] At the same time that man has protected them he has hunted out the original race which was "less useful to him,"[108] so that it is now to be found only in a few secluded spots, such as the mountains of Greece, Cyprus, and Sardinia. Buffon does not consider even the differences between sheep and goats to be sufficiently characteristic to warrant their being classed as different species.

"I shall never tire," he continues, "of repeating--seeing how important the matter is--that we must not form our opinions concerning nature, nor differentiate (différencier) her species, by a reference to minor special characteristics. And, again, that systems, far from having illustrated the history of animals, have, on the contrary, served rather to obscure it ... leading, as they do, to the creation of arbitrary species which nature knows nothing about; perpetually confounding real and hypothetical existences; giving us false ideas as to the very essence of species; uniting them and separating them without foundation or knowledge, and often without our having seen the animal with which we are dealing."[109]

First and Second Views of Nature.

The twelfth volume begins with a preface, entitled "A First View of Nature," from which I take the following:--

"What cannot Nature effect with such means at her disposal? She can do all except either create matter or destroy it. These two extremes of power the deity has reserved for himself only; creation and destruction are the attributes of his omnipotence. To alter and undo, to develop and to renew--these are powers which he has handed over to the charge of Nature."[110]

The thirteenth volume opens with a second view of nature. After describing what a man would have observed if he could have lived during many continuous ages, Buffon goes on to say:--

"And as the number, sustenance, and balance of power among species is constant, Nature would present ever the same appearance, and would be in all times and under all climates absolutely and relatively the same, if it were not her fashion to vary her individual forms as much as possible. The type of each species is founded in a mould of which the principal features have been cut in characters that are ineffaceable and eternally permanent, but all the accessory touches vary; no one individual is the exact facsimile of any other, and no species exists without a large number of varieties. In the human race on which the divine seal has been set most firmly, there are yet varieties of black and white, large and small races, the Patagonian, Hottentot, European, American, Negro, which, though all descended from a common father, nevertheless exhibit no very brotherly resemblance to one another."[111]

On an earlier page there is a passage which I may quote as showing Buffon to have not been without some--though very imperfect--perception of the fact which evidently made so deep an impression upon his successor, Dr. Erasmus Darwin. I refer to that continuity of life in successive generations, and that oneness of personality between parents and offspring, which is the only key that will make the phenomena of heredity intelligible.

"Man," he says, "and especially educated man, is no longer a single individual, but represents no small part of the human race in its entirety. He was the first to receive from his fathers the knowledge which their own ancestors had handed down to them. These, having discovered the divine art of fixing their thoughts so that they can transmit them to their posterity, become, as it were, one and the same people with their descendants (*se sont, pour ainsi dire, identifiés avec leur neveux*); while our descendants will in their turn be one and the same people with ourselves (*s'identifieront avec nous*). This reunion in a single person of the experience of many ages, throws back the boundaries of man's existence to the utmost limits of the past; he is no

meme

longer a single individual, limited as other beings are to the sensations and experiences of to-day. In place of the individual we have to deal, as it were, with the whole species."[112]

"Differences in exterior are nothing in comparison with those in interior parts. These last must be regarded as the causes, while the others are but the effects. The interior parts of living beings are the foundation of the plan of their design; this is their essential form, their real shape, their exterior is only the surface, or rather the drapery in which their true figure is enveloped. How often does not the study of comparative anatomy show us that two exteriors which differ widely conceal interiors absolutely like each other, and, on the contrary, that the smallest internal difference is accompanied by the most marked differences of outward appearance, changing as it does even the natural habits, faculties and attributes of the animal?"[113]

Apes and Monkeys.

The fourteenth volume is devoted to apes and monkeys, and to the chapter with which the volumes on quadrupeds are brought to a conclusion--a chapter for which perhaps the most important position in the whole work is thus assigned. It is very long, and is headed "On Descent with Modification" ("De la Dégénération des Animaux"). This is the chapter in which Buffon enters more fully into the "causes or means" of the transformation of species.

At the opening of the chapter on the nomenclature

of monkeys, the theory is broached that there is a certain fixed amount of life-substance as of matter in nature; and that neither can be either augmented or diminished. Buffon maintains this organic and living substance to be as real and durable as inanimate matter; as permanent in its state of life as the other in that of death; it is spread over the whole of nature, and passes from vegetables to animals by way of nutrition, and from animals back to vegetables through putrefaction, thus circulating incessantly to the animation of all that lives.

As might be expected, Buffon is loud in his protest against any real similarity between man and the apes--man has had the spirit of the Deity breathed into his nostrils, and the lowest creature with this is higher than the highest without it. Having settled this point, he makes it his business to show how little difference in other respects there is between the apes and man.

"One who could view," he writes, "Nature in her entirety, from first to last, and then reflect upon the manner in which these two substances--the living and the inanimate--act and react upon one another, would see that every living being is a mould which casts into its own shape those substances upon which it feeds; that it is this assimilation which constitutes the growth of the body, whose development is not simply an augmentation of volume, but an extension in all its dimensions, a penetration of new matter into all parts of its mass: he would see that these parts augment proportionately with the whole, and the whole proportionately with these parts, while general configuration remains the same until the full development is accomplished.... He would see that man, the quadruped, the cetacean, the bird, reptile, insect, tree, plant, herb, all are nourished, grow, and reproduce themselves on this same system, and that though their manner of feeding and of reproducing themselves may appear so different, this is only because the general and common cause upon which these operations depend can only operate in the individual agreeably with the form of each species. Travelling onward (for it has taken the human mind ages to arrive at these great truths, from which all others are derived), he would compare living forms, give them names to distinguish them, and other names to connect them with each other. Taking his own body as the model with which all living forms should be compared, and having measured them, explained them thoroughly, and compared them in all their parts, he would see that there is but small difference between the forms of living beings; that by dissecting the ape he could arrive at the anatomy of man, and that taking some other animal we find always the same ultimate plan of organization, the same senses, the same viscera, the same bones, the same flesh, the same movements of the fluids, the same play and action of the solids; he would find all of them with a heart, veins, arteries, in all the same organs of circulation, respiration, digestion, nutrition, secretion; in all of them a solid frame, composed of pieces put together in nearly the same manner; and he would find this system always the same, from man to the ape, from the ape to the quadrupeds, from the quadrupeds to the cetacea, birds, fishes, reptiles; this system or plan then, I say, if firmly laid hold of and comprehended by the human mind, is a true copy of nature; it is the simplest and most general point of view from which we can consider her, and if we extend our view, and go on from what lives to what vegetates, we may see this plan--which originally did but vary almost imperceptibly--change its scope and descend gradually from reptiles to insects, from insects to worms, from worms to zoophytes, from zoophytes to plants, and yet keeping ever the same fundamental unity in spite of differences of detail, insomuch that nutrition, development, and reproduction remain the common traits of all organic bodies; traits eternally essential and divinely implanted; which time, far from effacing or destroying, does but make plainer and plainer continually."

This is the writer who can see nothing in common between the horse and the zebra except that each has a solid hoof.[114] He continues:--

"If from this grand tableau of resemblances, in which the living universe presents itself to our eyes as though it were a single family, we pass to a tableau rather of the differences between living forms, we shall see that, with the exception of some of the greater species, such as the elephant, rhinoceros, hippopotamus, tiger, lion, which must each have their separate place, the other races seem all to blend with neighbouring forms, and to fall into groups of likenesses, greater or lesser, and of genera which our nomenclators represent to us by a network of shapes, of which some are held together by the feet, others by the teeth, horns, and skin, and others by points of still minor importance. And even those whose form strikes us as most perfect, as approaching most nearly to our own--even the apes--require some attention before they can be distinguished from one another, for the privilege of being an isolated species has been assigned less to form than to size; and man himself, though of a separate species and differing infinitely from all or any others, has but a medium size, and is less isolated and has nearer neighbours than have the greater animals. If we study the Orang-outang with regard only to his configuration, we might regard him, with equal justice, as either the highest of the apes or as the lowest of mankind, because, with the exception of the soul, he wants nothing of what we have ourselves, and because, as regards his body, he differs less from man than he does from other animals which are still called apes."[115]

The want of a soul Buffon maintains to be the only essential difference between the Orang-outang and man--"his body, limbs, senses, brain and tongue are the same as ours. He can execute whatever movements man can execute; yet he can neither think nor speak, nor do any action of a distinctly human character. Is this merely through want of training? or may it not be through wrong comparison on our own parts? We compare the wild ape in the woods to the civilized citizen of our great towns. No wonder the ape shows to disadvantage. He should be compared with the hideous Hottentot rather, who is himself almost as much above the lowest man, as the lowest man is above the Orang-outang."[116]

The passage is a much stronger one than I have

thought it fit to quote. The reader can refer to it for himself. After reading it I entertain no further doubt that Buffon intended to convey the impression that men and apes are descended from common ancestors. He was not, however, going to avow this conclusion openly.

"I admit," he continues, "that if we go by mere structure the ape might be taken for a variety of the human race; the Creator did not choose to model mankind upon an entirely distinct system from the other animals: He comprised their form and man's under a plan which is in the main uniform."[117] Buffon then dwells upon the possession of a soul by man; "even the lowest creature," he avers, "which had this, would have become man's rival."

"The ape then is purely an animal, far from being a variety of our own species, he does not even come first in the order of animals, since he is not the most intelligent: the high opinion which men have of the intelligence of apes is a prejudice based only upon the resemblance between their outward appearance and our own."[118] But the undiscerning were not only to be kept quiet, they were to be made happy. With this end, if I am not much mistaken, Buffon brings his chapter on the nomenclature of apes to the following conclusion:--

"The ape, which the philosopher and the uneducated have alike regarded as difficult to define, and as being at best equivocal, and midway between man and the lower animals, proves in fact to be an animal and nothing more; he is masked externally in the shape of man, but internally he is found incapable of thought, and of all that constitutes man; apes are below several of the other animals in respect of qualities corresponding to their own, and differ essentially from man, in nature, temperament, the time which must be spent upon their gestation and education, in their period of growth, duration of life, and in fact in all those profounder habits which constitute what is called the 'nature' of any individual existence."[119] This is handsome, and leaves the more timorous reader in full possession of the field.

Buffon is accordingly at liberty in the following chapter to bring together every fact he can lay his hands on which may point the resemblance between man and the Orang-outang most strongly; but he is careful to use inverted commas here much more freely than is his wont. Having thus made out a strong case for the near affinity between man and the Orang-outang, and having thrown the responsibility on the original authors of the passages he quotes, he excuses himself for having quoted them on the ground that "everything may seem important in the history of a brute which resembles man so nearly," and then insists upon the points of difference between the Orang-outang and ourselves. They do not, however, in Buffon's hands come to much, until the end of the chapter, when, after a *résumé* dwelling on the points of resemblance, the differences are again emphatically declared to have the best of it.

I need not follow Buffon through his description of the remaining monkeys. It comprises 250 pp., and is confined to details with which we have no concern;

but the last chapter--"De la Dégénération des Animaux"--deserves much fuller quotation than my space will allow me to make from it. The chapter is very long, comprising, as I have said, over sixty quarto pages. It is impossible, therefore, for me to give more than an outline of its contents.

Causes or Means of the Transformation of Species.

The human race is declared to be the one most capable of modification, all its different varieties being descended from a common stock, and owing their more superficial differences to changes of climate, while their profounder ones, such as woolly hair, flat noses, and thick lips, are due to differences of diet, which again will vary with the nature of the country inhabited by any race. Changes will be exceedingly gradual; it will take centuries of unbroken habit to bring about modifications which can be transmitted with certainty so as to eventuate in national characteristics.[120] It is a pleasure to find that here, too, habit is assigned as the main cause which underlies heredity.

Modification will be much prompter with animals. When compelled to abandon their native land, they undergo such rapid and profound modification, that at first sight they can hardly be recognized as the same race, and cannot be detected in their disguise till after the most careful inspection, and on grounds of analogy only. Domestication will produce still more surprising results; the stigmata of their captivity, the marks of their chains, can be seen upon all those animals which man has enslaved; the older and more confirmed the servitude, the deeper will be its scars, until at length it will be found impossible to rehabilitate the creature and restore to it its lost attributes.

"Temperature of climate, quality of food, and the ills of slavery--here are the three main causes of the alteration and degeneration of animals. The consequences of each of these should be particularly considered, so that by examining Nature as she is to-day we may thus perceive what she was in her original condition."[121]

I have more than once admitted that there is a wide difference between this opinion, which assigns modification to the direct influence of climate, food, and other changed conditions of life, and that of Dr. Erasmus Darwin, which assigns only an indirect effect to these, while the direct effect is given to changed actions in consequence of changed desires; but it is surprising how nearly Buffon has approached the later and truer theory, which may perhaps have been suggested to Dr. Darwin by the following pregnant passage--as pregnant, probably, to Buffon himself as to another:--

"The camel is the animal which seems to me to have felt the weight of slavery most profoundly. He is born with wens upon his back and callosities upon his knees and chest; these callosities are the unmistakable results of rubbing, for they are full of pus and of corrupted blood. The camel never walks without carrying a heavy burden, and the pressure of this has hindered, for generations, the free extension and uniform growth of the muscular parts of the back; whenever he reposes or sleeps his driver compels him to do so upon his folded legs, so that little by little this position becomes habitual with him. All the weight of his body bears, during several hours of the day continuously, upon his chest and knees, so that the skin of these parts, pressed and rubbed against the earth, loses its hair, becomes bruised, hardened, and disorganized.

"The llama, which like the camel passes its life beneath burdens, and also reposes only by resting its weight upon its chest, has similar callosities, which again are perpetuated in successive generations. Baboons, and pouched monkeys, whose ordinary position is a sitting one, whether waking or sleeping, have callosities under the region of the haunches, and this hard skin has even become inseparable from the bone against which it is being continually pressed by the weight of the body; in the case, however, of these animals the callosities are dry and healthy, for they do not come from the constraint of trammels, nor from the burden of a foreign weight, but are the effects only of the natural habits of the animal, which cause it to continue longer seated than in any other position. There are callosities of these pouched monkeys which resemble the double sole of skin which we have ourselves under our feet; this sole is a natural hardness which our continued habit of walking or standing upright will make thicker or thinner according to the greater or less degree of friction to which we subject our feet."[122]

This involves the whole theory of Dr. Darwin.

Wild animals would not change either their food or

climate if left to themselves, and in this case they would not vary, but either man or some other enemies have harassed most of them into migrations; "those whose nature was sufficiently flexible to lend itself to the new situation spread far and wide, while others have had no resource but the deserts in the neighbourhood of their own countries."[123]

Since food and climate, and still less man's empire over them, can have but little effect upon wild animals, Buffon refers their principal varieties in great measure to their sexual habits, variations being much less frequent among animals that pair and breed slowly, than among those which do not mate and breed more freely. After running rapidly over several animals, and discussing the flexibility or inflexibility of their organizations, he declares the elephant to be the only one on which a state of domestication has produced no effect, inasmuch as "it refuses to breed under confinement, and cannot therefore transmit the badges of its servitude to its descendants."[124]

Here is an example of Buffon's covert manner, in the way he maintains that descent with modification may account not only for specific but for generic differences.

"But after having taken a rapid survey of the varieties which indicate to us the alterations that each species has undergone, there arises a broader and more important question, how far, namely, species themselves can change--how far there has been an older degeneration, immemorial from all antiquity, which has taken place in every family, or, if the term is preferred, *in all the genera* under which those species are comprehended which neighbour one another without presenting points of any very profound dissimilarity? We have only a few isolated species, such as man, which form at once the species and the whole genus; the elephant, the rhinoceros, the hippopotamus, and the giraffe form genera, or simple species, which go down in a single line, with no collateral branches. All other races appear to form families, in which we may perceive a common source or stock from which the different branches seem to have sprung in greater or less numbers according as the individuals of each species are smaller and more fecund."[125]

I can see no explanation of the introduction of this passage unless that it is intended to raise the question whether modification may be not only specific but generic, the point of the paragraph lying in the words "dans chaque famille, *ou si l'on veut, dans chacun des genres*." We are told in the next paragraph, that if we choose to look at the matter in this light, well--in that case--we ought to see not only the ass and the horse, but *the zebra too*, as members of the same family; "the number of their points of resemblance being infinitely greater than those in respect of which they differ."[126] Thus, at the close of his work on the quadrupeds, he thinks it well, as at the commencement seventeen years earlier, to emphasize--in his own quiet way--his perception that the principles on which he has been insisting should be carried much farther than he has chosen to carry them.

His conclusion is, that "after comparing all the animals and bringing them each under their proper genus, we shall find the two hundred species we have already described to be reducible into a sufficiently small number of families or main stocks from which it is not impossible that all the others may be derived."[127]

The chapter closes thus:--

"To account for the origin of these animals" (certain of those peculiar to America), "we must go back to the time when the two continents were not yet separated, and call to mind the earliest geological changes. At the same time, we must consider the two hundred existing species of quadrupeds as reduced to thirty-eight families. And though this is not at all the state of Nature as she is in our time, and as she has been represented in this volume, and though, in fact, it is a condition which we can only arrive at by induction, and by analogies almost as difficult to lay hold of as is the time which has effaced the greater number of their traces, I shall, nevertheless, endeavour to ascend to these first ages of Nature by the aid of facts and monuments which yet remain to us, and to represent the epochs which these facts seem to indicate."[128]

The fifteenth volume contains a description of a few more monkeys, as also of some animals which Buffon had never actually seen, a great part being devoted to indices.

Supplement.

The first four volumes of the Supplement to Buffon's 'Natural History,' 1774-1789, contain little which throws additional light upon his opinions concerning the mutability of species. At the beginning, however, of the fifth volume I find the following:--

"On comparing these ancient records of the first ages of life [fossils] with the productions of to-day, we see with sufficient clearness that the essential form has been preserved without alteration in its principal parts: there has been no change whatever in the general type of each species; the plan of the inner parts has been preserved without variation. However long a time we may imagine for the succession of ages, whatever number of generations we may suppose, the individuals of to-day present to us in each genus the same forms as they did in the earliest ages; and this is more especially true of the greater species, whose characters are more invariable and nature more fixed; for the inferior species have, as we have said, experienced in a perceptible manner all the effects of different causes of degeneration. Only it should be remarked in regard to these greater species, such as the elephant and hippopotamus, that in comparing their fossil remains with the existing forms we find the earlier ones to have been larger. Nature was then in the full vigour of her youth, and the interior heat of the earth gave to her productions all the force and all the extent of which they were capable ... if there have been lost species, that is to say animals which existed once, but no longer do so, these can only have been animals which required a heat greater than that of our present torrid zone."[129]

The context proves Buffon to have been thinking of such huge creatures as the megatherium and mastodon, but his words seem to limit the extinction of species to the denizens of a hot climate which had turned colder. It is not at all likely that Buffon meant this, as the passage quoted at p. 146 of this work will suffice to show. The whole paragraph is ironical.

I can see nothing to justify the conclusion drawn from this passage by Isidore Geoffroy, that Buffon had modified his opinions, and was inclined to believe in a more limited mutability than he had done a few years earlier. His exoteric position is still identical with what it was in the outset, and his esoteric may be seen from the spirit which is hardly concealed under the following:--

"I shall be told that analogy points towards the belief that our own race has followed the same path, and dates from the same period as other species; that it has spread itself even more widely than they; and that if man's creation has a later date than that of the other animals, nothing shows that he has not been subjected to the same laws of nature, the same alterations, and the same changes as they. We will grant that the human species does not differ essentially from others in the matter of bodily organs, and that, in respect of these, our lot has been much the same as that of other animals."[130]

Plants under Domestication.

"If more modern and even recent examples are required in order to prove man's power over the vegetable kingdom, it is only necessary to compare our vegetables, flowers, and fruits with the same species such as they were a hundred and fifty years ago; this can be done with much ease and certainty by running the eye over the great collection of coloured drawings begun in the time of Gaston of Orleans, and continued to the present day at the Jardin du Roi. We find with surprise that the finest flowers of that date, as the ranunculuses, pinks, tulips, bear's ears, &c., would be rejected now, I do not say by our florists, but by our village gardeners. These flowers, though then already cultivated, were still not far above their wild condition. They had a single row of petals only, long pistils, colours hard and false; they had little velvety texture, variety, or gradation of tints, and, in fact, presented all the characteristics of untamed nature. Of herbs there was a single kind of endive, and two of lettuce--both bad--while we can now reckon more than fifty lettuces and endives, all excellent. We can even name the very recent dates of our best pippins and kernel fruits--all of them differing from those of our forefathers, which they resemble in name only. In most cases things remain while names change; here, on the contrary, it is the names that have been constant while the things have varied.[131]

"It is not that every one of these good varieties did not arise from the same wild stock; but how many attempts has not man made on Nature before he succeeded in getting them. How many millions of germs has he not committed to the earth, before she has rewarded him by producing them? It was only by sowing, tending, and bringing to maturity an almost infinite number of plants of the same kind that he was able to recognize some individuals with fruits sweeter and better than others; and this first discovery, which itself involves so much care, would have remained for ever fruitless if he had not made a second, which required as much genius as the first required patience--I mean the art of grafting those precious individuals, which, unfortunately, cannot continue a line as noble as their own, nor themselves propagate their rare and admirable qualities? And this alone proves that these qualities are purely individual, and not specific, for the pips or stones of these excellent fruits bring forth the original wild stock, so that they do not form species essentially different from this. Man, however, by means of grafting, produces what may be called secondary species, which he can propagate at will; for the bud or small branch which he engrafts upon the stock contains within itself the individual quality which cannot be transmitted by seed, but which needs only to be developed in order to bring forth the same fruits as the individual from which it was taken in order to be grafted on to the wild stock. The wild stock imparts none of its bad qualities to the bud, for it did not contribute to the forming thereof, being, as it were, a wet nurse, and no true mother.

"In the case of animals, the greater number of those features which appear individual, do not fail to be transmitted to offspring, in the same way as specific characters. It was easier then for man to produce an effect upon the natures of animals than of plants. The different breeds in each animal species are variations that have become constant and hereditary, while vegetable species on the other hand present no variations that can be depended on to be transmitted with certainty.

"In the species of the fowl and the pigeon alone, a large number of breeds have been formed quite recently, which are all constant, and in other species we daily improve breeds by crossing them. From time to time we acclimatize and domesticate some foreign and wild species. All these examples of modern times prove that man has but tardily discovered the extent of his own power, and that he is not even yet sufficiently aware of it. It depends entirely upon the exercise of his intelligence; the more, therefore, he observes and cultivates nature the more means he will find of making her subservient to him, and of drawing new riches from her bosom without diminishing the treasures of her inexhaustible fecundity."[132]

Birds.

In the preface to his volumes upon birds, Buffon says that these are not only much more numerous than quadrupeds, but that they also exhibit a far larger number of varieties, and individual variations.

"The diversities," he declares, "which arise from the effects of climate and food, of domestication, captivity, transportation, voluntary and compulsory migration--all the causes in fact of alteration and degeneration--unite to throw difficulties in the way of the ornithologist."[133]

He points out the infinitely keener vision of birds than that of man and quadrupeds, and connects it with their habits and requirements.[134] He does not appear to consider it as caused by those requirements, though it is quite conceivable that he saw this, but thought he had already said enough. He repeatedly refers to the effects of changed climate and of domestication, but I find nothing in the first volume which modifies the position already taken by him in regard to descent with modification: it is needless, therefore, to repeat the few passages which are to be found bearing at all upon the subject. The chapter on the birds that cannot fly, contains a sentence which seems to be the germ that has been developed, in the hands of Lamarck, into the comparison between nature and a tree. Buffon says that the chain of nature is not a single long chain, but is comparable rather to something woven, "which at certain intervals throws out a branch sideways that unites it with the strands of some other weft."[135] On the following page there is a passage which has been quoted as an example of Buffon's contempt for the men of science of his time. The writer maintains that the most lucid arrangement of birds, would have been to begin with those which most resembled quadrupeds. "The ostrich, which approaches the camel in the shape of its legs, and the porcupine in the quills with which its wings are armed, should have immediately followed the quadrupeds, but philosophy is often obliged to make a show of yielding to popular opinions, and *the tribe of naturalists* is both numerous and impatient of any disturbance of its methods. It would only, then, have regarded this arrangement as an unreasonable innovation caused by a desire to contradict and to be singular."[136]

It is, I believe, held not only by "*le peuple des naturalistes*," but by most sensible persons, that the proposed arrangement would not have been an improvement. I find, however, in the preface to the third volume on birds that M. Gueneau de Montbeillard described all the birds from the ostrich to the quail, so the foregoing passage is perhaps his and not Buffon's. If so, the imitation is fair, but when we reflect upon it we feel uncertain whether it is or is not beneath Buffon's dignity.

Here, as often with pictures and music, we cannot criticise justly without taking more into consideration than is actually before us. We feel almost inclined to say that if the passage is by Buffon it is probably right, and if by M. Gueneau de Montbeillard, probably wrong. It must also be remembered that, as we learn from the preface already referred to, Buffon was seized at this point in his work with a long and painful illness, which continued for two years; a single hasty passage in so great a writer may well be pardoned under such circumstances.

Looking through the third and remaining volumes on birds, the greater part of which was by Gueneau de Montbeillard, and bearing in mind that in point of date they are synchronous with some of those upon quadrupeds from which I have already extracted as much as my space will allow, and not seeing anything on a rapid survey which promises to throw new light upon the author's opinions, I forbear to quote further. I therefore leave Buffon with the hope that I have seen him more justly than some others have done, but with the certainty that the points I have caught and understood are few in comparison with those that I have missed.

FOOTNOTES:

[65] 'Hist. Nat.,' tom. i. p. 13, 1749.

[66] Ibid.

[67] Ibid. p. 16.

[68] Tom. i. p. 21.

[69] Ibid. p. 23.

[70] Tom. ii. p. 9, 1749.

[71] Ibid. p. 10.

[72] Tom. iv. p. 31, 1753.

[73] Tom. iv. p. 55.

[74] Tom. iv. p. 98, 1753.

[75] Ibid.

[76] Tom. viii. p. 283, &c., 1760.

[77] Tom. iv. p. 102, 1760.

[78] Tom. iv. p. 103, 1753.

[79] Dr. Darwin, 'Zoonomia,' vol. i. p. 183, 1796.

[80] Ibid. p. 184.

[81] Dr. Darwin,'Zoonomia,' vol. i. p. 186.

[82] Tom. v. p. 63, 1755.

[83] Ibid. p. 64.

[84] Tom. v. p. 103, 1755.

[85] Tom. v. p. 104, 1755.

[86] Tom. v. pp. 192-195, 1755.

[87] Tom. v. p. 195.

[88] Tom. v. pp. 196, 197.

[89] This passage would seem to be the one which has suggested the following to the author of 'The Vestiges of Creation':--

"He [the Deity] has endowed the families which enjoy His bounty with an almost infinite fecundity, ... but the limitation of the results of this fecundity ... is accomplished in a befitting manner by His ordaining that certain other animals shall have endowments sure so to act as to bring the rest of animated beings to a proper balance" (p. 317, ed. 1853).

[90] Tom. vi. p. 252, 1756.

[91] 'Discours sur la Nature des Animaux,' vol. iv. and p. 113 of this vol.

[92] Tom. vii. p. 9, 1758.

[93] Tom. vii. p. 10, 1758.

[94] Tom. vii. p. 12, 1758.

[95] Tom. vii. p. 14, 1758

[96] Tom. vii. p. 15, 1758.

[97] Tom. vii. p. 19, 1758.

[98] Tom. vii. p. 23, 1758. See Sténon's Discourse upon this subject.

[99] Tom. ix. p. 10, 1761.

[100] Tom. ix. p. 11, 1761.

[101] Tom. ix. p. 68, 1761.

[102] Ibid. p. 96, 1761.

[103] Tom. ix. p. 107 and following pages (during which he rails at the new world generally), 1761.

[104] Tom. ix. p. 127, 1761.

[105] Tom. xi. p. 290, 1764 (misprinted on title-page 1754).

[106] Ibid. p. 296.

[107] Ibid. p. 363.

[108] Ibid. p. 363.

[109] Tom. xi. p. 370, 1764.

[110] Ibid. xii., preface, iv. 1764.

[111] Tom. xiii., preface, x. 1765.

[112] Tom. xiii., preface, iv. 1765.

[113] Ibid. xiii. p. 37.

[114] See p. 80 of this volume.

[115] Tom. xiv. p. 30, 1766.

[116] Tom. xiv. p. 31, 1766.

[117] Ibid. p. 32, 1766.

[118] Tom. xiv. p. 38, 1766.

[119] Ibid. p. 42, 1766.

[120] Tom. xiv. p. 316, 1766.

[121] Ibid. p. 317.

[122] Tom. xiv. p. 326, 1766.

[123] Ibid. p. 327.

[124] Tom. xiv. p. 333.

[125] Ibid. p. 335, 1766.

[126] See p. 80 of this volume.

[127] Tom. xiv. p. 358, 1766.

[128] Tom. xiv. p. 374, 1766.

[129] 'Hist. Nat.,' Sup. tom. v. p. 27, 1778.

[130] Sup. tom. v. p. 187, 1778.

[131] Sup. tom. v. p. 250, 1778.

[132] Sup. tom. v. p. 253, 1778.

[133] 'Oiseaux,' tom. i., preface, v. 1770.

[134] Ibid. pp. 9-11.

[135] 'Oiseaux,' tom. i. pp. 394, 395.

[136] Ibid. p. 396, 1771.

CHAPTER XII.

SKETCH OF DR. ERASMUS DARWIN'S LIFE.

Proceeding now to the second of the three founders of the theory of evolution, I find, from a memoir by Dr. Dowson, that Dr. Erasmus Darwin was born at Elston, near Newark, in Nottinghamshire, on the 12th of December, 1731, being the seventh child and fourth son of Robert Darwin, "a private gentleman, who had a taste for literature and science, which he endeavoured to impart to his sons. Erasmus received his early education at Chesterfield School, and later on was entered at St. John's College, Cambridge, where he obtained a scholarship of about 16*l.* a year, and distinguished himself by his poetical exercises, which he composed with uncommon facility. He took the degree of M.B. there in 1755, and afterwards prepared himself for the practice of medicine by attendance on the lectures of Dr. Hunter in London, and a course of studies in Edinburgh.

"He first settled as a physician at Nottingham; but meeting with no success there, he removed in the autumn of 1756, his twenty-fifth year, to Lichfield, where he was more fortunate; for a few weeks after his arrival, to use the words of Miss Seward, 'he brilliantly opened his career of fame.' A young gentleman of family and fortune lay sick of a dangerous fever. A physician who had for many years possessed the confidence of Lichfield and the neighbourhood attended, but at length pronounced the case hopeless, and took his leave. Dr. Darwin was then called in, and by 'a reverse and entirely novel kind of treatment' the patient recovered."[137]

Of Dr. Darwin's personal appearance Miss Seward says:--

"He was somewhat above the middle size; his form athletic, and inclined to corpulence; his limbs were too heavy for exact proportion; the traces of a severe smallpox disfigured features and a countenance which, when they were not animated by social pleasure, were rather saturnine than sprightly; a stoop in the shoulders, and the then professional appendage--a large full-bottomed wig--gave at that early period of life an appearance of nearly twice the years he bore. Florid health and the earnest of good humour, a funny smile on entering a room and on first accosting his friends, rendered in his youth that exterior agreeable, to which beauty and symmetry had not been propitious.

"He stammered extremely, but whatever he said, whether gravely or in jest, was always well worth waiting for, though the inevitable impression it made might not be always pleasant to individual self-love. Conscious of great native elevation above the general standard of intellect, he became early in life sore upon opposition, whether in argument or conduct, and always resented it by sarcasm of very keen edge. Nor was he less impatient of the sallies of egotism and vanity, even when they were in so slight a degree that strict politeness would rather tolerate than ridicule them. Dr. Darwin seldom failed to present their caricature in jocose but wounding irony. If these ingredients of colloquial despotism were discernible in *unworn* existence, they increased as it advanced, fed by an ever growing reputation within and without the pale of medicine."[138]

I imagine that this portrait is somewhat too harshly drawn. Dr. Darwin's taste for English wines is the worst trait which I have been able to discover in his character. On this head Miss Seward tells us that "he despised the prejudice which deems foreign wines more wholesome than the wines of the country. 'If you must drink wine,' said he, 'let it be home-made.'" "It is well known," she continues, "that Dr. Darwin's influence and example have sobered the county of Derby; that intemperance in fermented fluid of every species is almost unknown among its gentlemen,"[139] which, if he limited them to cowslip wine, is hardly to be wondered at.

Dr. Dowson, quoting Miss Edgeworth, says that Dr. Darwin attributed almost all the diseases of the upper classes to the too great use of fermented liquors. "This opinion he supported in his writings with the force of his eloquence and reason; and still more in conversation by all those powers of wit, satire, and peculiar humour, which never appeared fully to the public in his works, but which gained him strong ascendancy in private society.... When he heard that my father was bilious, he suspected that this must be the consequence of his having, since his residence in Ireland, and in compliance with the fashion of the country, indulged too freely in drinking. His letter, I remember, concluded with, 'Farewell, my dear friend; God keep you from whisky--if He can.'"[140]

On the other hand, Dr. Darwin seems to have been a very large eater. "Acid fruits with sugar, and all sorts of creams and butter were his luxuries; but he always ate plentifully of animal food. This liberal alimentary regimen he prescribed to people of every age where unvitiated appetite rendered them capable of following it; even to infants."

Dr. Dowson writes:--

"I have mentioned already that he had in his carriage a receptacle for paper and pencils, with which he wrote as he travelled, and in one corner a pile of books; but he had also a receptacle for a knife, fork, and spoon, and in the other corner a hamper, containing fruit and sweetmeats, cream and sugar. He provided also for his horses by having a large pail lashed to his carriage for watering them, as well as hay and oats to be eaten on the road. Mrs. Schimmelpenninck says that when he came on a professional visit to her father's house they had, as was the custom whenever he came, 'a luncheon-table set out with hothouse fruits and West India sweetmeats, clotted cream, stilton cheese, &c. While the conversation went on, the dishes in his vicinity were rapidly emptied, and what,' she adds, 'was my astonishment when, at the end of the three hours during which the meal had lasted, he expressed his joy at hearing the dressing bell, and hoped dinner would soon be announced.' This was not mere gluttony; he thought an abundance, or what most people would consider a superabundance of food, conducive to health. '*Eat or be eaten*' is said to have been often his medical advice. He had especially a very high opinion of the nutritive value of sugar, and said 'that if ever our improved chemistry should discover the art of making sugar from fossil or aerial matter without the assistance of vegetation, food for animals would then become as plentiful as water, and mankind might live upon the earth as thick as blades of grass, with no restraint to their numbers but want of room.'--Botanic Garden, vol. i. p. 470."[141]

"Professional generosity," says Miss Seward, "distinguished Dr. Darwin's practice. Whilst resident in Lichfield he always cheerfully gave to the priest and lay vicars of its cathedral and their families *his advice*, but never took fees from any of them. Diligently also did he attend the health of the poor in that city, and afterwards at Derby, and supplied their necessities by food, and all sort of charitable assistance. In each of those towns *his* was the cheerful board of almost open-housed hospitality, without extravagance or parade; generosity, wit, and science were his household gods."[142]

Of his first marriage the following account is given:--

"In 1757 he married Miss Howard, of the Close of Lichfield, a blooming and lovely young lady of eighteen.... Mrs. Darwin's own mind, by nature so well endowed, strengthened and expanded in the friendship, conversation, and confidence of so beloved a preceptor. But alas! upon her too early youth, and too delicate constitution, the frequency of her maternal situation, during the first five years of her marriage, had probably a baneful effect. The potent skill and assiduous cares of *him* before whom disease daily vanished from the frame of *others*, could not expel it radically from that of her he loved. It was, however, kept at bay during thirteen years.

"Upon the distinguished happiness of those years she spoke with fervour to two intimate female friends in the last week of her existence, which closed at the latter end of the summer 1770. 'Do not weep for my impending fate,' said the dying angel with a smile of unaffected cheerfulness. 'In the short term of my life a great deal of happiness has been comprised. The maladies of my frame were peculiar; those of my head and stomach which no medicine could eradicate, were spasmodic and violent; and required stronger measures to render them supportable while they lasted than my constitution could sustain without injury. The periods of exemption from those pains were frequently of several days' duration, and in my intermissions I felt no indications of malady. Pain taught me the value of ease, and I enjoyed it with a glow of spirit, seldom, perhaps, felt by the habitually healthy. While Dr. Darwin combated and assuaged my disease from time to time, his indulgence to all my wishes, his active desire to see me amused and happy, proved incessant. His house, as you know, has ever been the resort of people of science and merit. If, from my husband's great and extensive practice, I had much less of his society than I wished, yet the conversation of his friends, and of my own, was ever ready to enliven the hours of his absence. As occasional malady made me doubly enjoy health, so did those frequent absences give a zest even to delight, when I could be indulged with his company. My three boys have ever been docile and affectionate. Children as they are, I could trust them with important secrets, so sacred do they hold every promise they make. They scorn deceit and falsehood of every kind, and have less selfishness than generally belongs to childhood. Married to any other man, I do not suppose I could have lived a third part of the years which I have passed with Dr. Darwin; he has prolonged my days, and he has blessed them.'

"Thus died this superior woman, in the bloom of life, sincerely regretted by all who knew how to value her excellence, and *passionately* regretted by the selected few whom she honoured with her personal and confidential friendship."[143]

I find Miss Seward's pages so fascinating, that I am in danger of following her even in those parts of her work which have no bearing on Dr. Darwin. I must, however, pass over her account of Mr. Edgeworth and of his friend Mr. Day, the author of 'Sandford and Merton,' "which, by wise parents, is put into every youthful hand," but the description of Mr. Day's portrait cannot be omitted.

"In the course of the year 1770, Mr. Day stood for a full-length picture to Mr. Wright, of Derby. A strong likeness and a dignified portrait were the result. Drawn in the open air, the surrounding sky is tempestuous, lurid, dark. He stands leaning his left arm against a column inscribed to Hambden (*sic*). Mr. Day looks upwards, as enthusiastically meditating on the contents of a book held in his dropped right hand. The open leaf is the oration of that virtuous patriot in the senate, against the grant of ship money, demanded by King Charles I. A flash of lightning plays in Mr. Day's hair, and illuminates the contents of the volume. The poetic fancy and what were *then* the politics of the original, appear in the choice of subject and attitude. Dr. Darwin sat to Mr. Wright about the same period. *That* was a simply contemplative portrait, of the most perfect resemblance."[144]

"In the year 1768, Dr. Darwin met with an accident of irretrievable injury to the human frame. His propensity to mechanics had unfortunately led him to construct a very singular carriage. It was a platform with a seat fixed upon a very high pair of wheels, and supported in the front upon the back of the horse, by means of a kind of proboscis which, forming an arch, reached over the hind-quarters of the horse, and passed through a ring, placed on an upright piece of iron, which worked in a socket fixed in the saddle. The horse could thus move from one side of the road to the other, quartering, as it is called, at the will of the driver, whose constant attention was necessarily employed to regulate a piece of machinery contrived, but *not well* contrived, for that purpose."

I cannot help the reader to understand the foregoing description. "From this whimsical carriage, however, the doctor was several times thrown, and the last time he used it had the misfortune, from a similar accident, to break the patella of his right knee, which caused, as it must always cause, an incurable weakness in the fractured part, and a lameness not very discernible, indeed, when walking on even ground."[145]

Miss Seward presently tells a story which reads as though it might have been told by Plutarch of some Greek or Roman sage. Much as we must approve of Dr. Darwin's habitual sobriety, we shall most of us be agreed that a few more such stories would have been cheaply purchased by a corresponding number of lapses on the doctor's part.

Miss Seward writes:--

"Since these memoirs commenced, an odd anecdote of Dr. Darwin's early residence at Lichfield, was narrated to a friend of the author by a gentleman, who was of the party in which it happened. Mr. Sneyd, then of Bishton, and a few more gentlemen of Staffordshire, prevailed upon the doctor to join them in an expedition by water from Burton to Nottingham, and on to Newark. They had cold provisions on board, and plenty of wine. It was midsummer; the day ardent and sultry. The noon-tide meal had been made, and the glass had gone gaily round. It was one of those *few* instances in which the medical votary of the Naiads transgressed his general and strict sobriety," in which, in fact, he may be said to have--remembered himself.

"If not absolutely intoxicated, his spirits were in a high state of vinous exhilaration. On the boat approaching Nottingham, within the distance of a few fields, he surprised his companions by stepping, without any previous notice, from the boat into the middle of the river, and swimming to shore. They saw him get upon the bank, and walk coolly over the meadows towards the town: they called to him in vain, but he did not once turn his head.

"Anxious lest he should take a dangerous cold by remaining in his wet clothes, and uncertain whether or not he intended to desert the party, they rowed instantly to the town at which they had not designed to have touched, and went in search of their river-god.

"In passing through the market-place they saw him standing upon a tub, encircled by a crowd of people, and resisting the entreaties of an apothecary of the place, one of his old acquaintances, who was importuning him to his house, and to accept other raiments till his own could be dried.

"The party on pressing through the crowd were surprised to hear him speaking without any degree of his usual stammer:--'Have I not told you, my friend, that I had drank a considerable quantity of wine before I committed myself to the river. You know my general sobriety, and as a professional man you *ought* to know that the *unusual* existence of internal stimulus would, in its effects upon the system, counteract the *external* cold and moisture.'"

"Then perceiving his companions near him, he nodded, smiled, and waived his hand, as enjoining them silence, thus, without hesitation, addressing the populace:--

"'Ye men of Nottingham, listen to me. You are ingenious and industrious mechanics. By your industry life's comforts are procured for yourselves and families. If you lose your health the power of being industrious will forsake you. *That* you know, but you may *not* know that to breathe fresh and changed air constantly, is not less necessary to preserve health than sobriety itself. Air becomes unwholesome in a few hours if the windows are shut. Open those of your sleeping rooms whenever you quit them to go to your workshops. Keep the windows of your workshops open whenever the weather is not insupportably cold. I have no *interest* in giving you this advice; remember what I, your countryman and a physician, tell you. If you would not bring infection and disease upon yourselves, and to your wives and little ones, change the air you breathe, change it many times a day, by opening your windows.'

"So saying, he stepped down from the tub, and, returning with his party to their boat, they pursued their voyage."[146]

Could any missionary be more perfectly sober and sensible, or more alive to the immorality of trying to effect too sudden a modification in the organisms he was endeavouring to influence? If the men of Nottingham want a statue in their market-place, I would respectfully suggest that a subject is here afforded them.

"Dr. Johnson was several times at Lichfield on visits to Mrs. Lucy Porter, his daughter-in-law, while Dr. Darwin was one of the inhabitants. They had one or two interviews, but never afterwards sought each other. Mutual and strong dislike subsisted between them. It is curious that in Johnson's various letters to Mrs. Thrale, now Mrs. Piozzi, published by that lady after his death, many of them dated from Lichfield, the name of Darwin cannot be found, nor, indeed, that of any of the ingenious and lettered people who lived there; while of its mere common-life characters there is frequent mention, with many hints of Lichfield's intellectual barrenness, while it could boast a Darwin and other men of classical learning, poetic talents, and liberal information."[147]

Here there follows a pleasant sketch of the principal Lichfield notabilities, which I am compelled to omit.

"*These* were the men," exclaims Miss Seward, "whose intellectual existence passed unnoticed by Dr. Johnson in his depreciating estimate of Lichfield talents. But Johnson liked only *worshippers*. Archdeacon Vyse, Mr. Seward, and Mr. Robinson paid all the respect and attention to Dr. Johnson, on these his visits to their town, due to his great abilities, his high reputation, and to whatever was estimable in his *mixed* character; but they were not in the herd that 'paged his heels,' and sunk in servile silence under the force of his dogmas, when their hearts and their judgments bore *contrary* testimony.

"Certainly, however, it was an arduous hazard to the feelings of the company to oppose in the slightest degree Dr. Johnson's opinions. His stentor lungs; that combination of wit, humour, and eloquence, which 'could make the *worse* appear the *better* reason,' that sarcastic contempt of his antagonist, never suppressed or even softened by the due restraints of good breeding, were sufficient to close the lips in his presence, of men who could have met him in fair argument, on *any* ground, literary or political, moral or characteristic.

"Where Dr. Johnson was, Dr. Darwin had no chance of being heard, though at least his equal in genius, his superior in science; nor, indeed, from his impeded utterance, in the company of any overbearing declaimer; and he was too intellectually great to be an humble listener to Johnson. Therefore he shunned him on having experienced what manner of man he was. The surly dictator felt the mortification, and revenged it by *affecting* to avow his disdain of powers too distinguished to be objects of *genuine* scorn.

"Dr. Darwin, in his turn, was not much more just to Dr. Johnson's genius. He uniformly spoke of him in terms which, had they been deserved, would have justified Churchill's 'immane Pomposo' as an appellation of *scorn*; since if his person was huge, and his manners pompous and violent, so were his talents vast and powerful, in a degree from which only prejudice and resentment could withhold respect.

"Though Dr. Darwin's hesitation in speaking precluded his flow of colloquial eloquence, it did not impede, or at all lessen, the force of that conciser quality, *wit*. Of satiric wit he possessed a very peculiar species. It was neither the dead-doing broadside of Dr. Johnson's satire, nor the aurora borealis of Gray ... whose arch yet coy and quiet fastidiousness of taste and feeling, as recorded by Mason, glanced bright and cold through his conversation, while it seemed difficult to define its nature; and while its effects were rather *perceived* than *felt*, exciting surprise more than mirth, and never awakening the pained sense of being the object of its ridicule. That unique in humorous verse, the Long Story, is a complete and beautiful specimen of Gray's singular vein.

"Darwinian wit is not more easy to be defined; instances will best convey an idea of its character to those who never conversed with its possessor.

"Dr. Darwin was conversing with a brother botanist concerning the plant kalmia, then a just imported stranger in our greenhouses and gardens. A lady who was present, concluding he had seen it, which in fact he had not, asked the doctor what were the colours of the plant. He replied, 'Madam, the kalmia has precisely the colours of a seraph's wing.' So fancifully did he express his want of consciousness concerning the appearance of a flower, whose name and rareness were all he knew of the matter.

"Dr. Darwin had a large company at tea. His servant announced a stranger, lady and gentleman. The female was a conspicuous figure, ruddy, corpulent, and tall. She held by the arm a little, meek-looking, pale, effeminate man, who, from his close adherence to the side of the lady, seemed to consider himself as under her protection.

"'Dr. Darwin, I seek you not as a physician, but as a

Belle Esprit. I make this husband of mine,' and she looked down with a side glance upon the animal, 'treat me every summer with a tour through one of the British counties, to explore whatever it contains worth the attention of ingenious people. On arriving at the several inns in our route I always search out the man of the vicinity most distinguished for his genius and taste, and introduce myself, that he may direct as the objects of our examination, whatever is curious in nature, art, or science. Lichfield will be our headquarters during several days. Come, doctor, whither must we go; what must we investigate to-morrow, and the next day, and the next? Here are my tablets and pencil.'

"'You arrive, madam, at a fortunate juncture. To-morrow you will have an opportunity of surveying an annual exhibition perfectly worthy your attention. To-morrow, madam, you will go to Tutbury bull-running.'

"The satiric laugh with which he stammered out the last word more keenly pointed this sly, yet broad rebuke to the vanity and arrogance of her speech. She had been up amongst the boughs, and little expected they would break under her so suddenly, and with so little mercy. Her large features swelled, and her eyes flashed with anger--'I was recommended to a man of genius, and I find him insolent and ill-bred.' Then, gathering up her meek and alarmed husband, whom she had loosed when she first spoke, under the shadow of her broad arm and shoulder, she strutted out of the room.

"After the departure of this curious couple, his

guests told their host he had been very unmerciful. 'I chose,' replied he, 'to avenge the cause of the little man, whose nothingness was so ostentatiously displayed by his lady-wife. Her vanity has had a smart emetic. If it abates the symptoms, she will have reason to thank her physician who administered without hope of a fee.'"[148]

"In the spring of 1778 the children of Colonel and Mrs. Pole of Radburn, in Derbyshire, had been injured by a dangerous quantity of the cicuta, injudiciously administered to them in the hooping-cough by a physician of the neighbourhood. Mrs. Pole brought them to the house of Dr. Darwin in Lichfield, remaining with them there a few weeks, till by his art the poison was expelled from their constitutions and their health restored.

"Mrs. Pole was then in the full bloom of her youth and beauty. Agreeable features; the glow of health; a fine form, tall and graceful; playful sprightliness of manner; a benevolent heart, and maternal affection, in all its unwearied cares and touching tenderness, contributed to inspire Dr. Darwin's admiration, and to secure his esteem."[149]

"In the autumn of this year" (1778) "Mrs. Pole of Radburn was taken ill; her disorder a violent fever. Dr. Darwin was called in, and never perhaps since the death of Mrs. Darwin, prescribed with such deep anxiety. Not being requested to continue in the house during the ensuing night, which he apprehended might prove critical, he passed the remaining hours till day-dawn beneath a tree opposite her apartment, watching the passing and repassing lights in the chamber. During the period in which a life so passionately valued was in danger, he paraphrased Petrarch's celebrated sonnet, narrating a dream whose prophecy was accomplished by the death of Laura. It took place the night on which the vision arose amid his slumber. Dr. Darwin extended the thought of that sonnet into the following elegy:--

"Dread dream, that, hovering in the midnight air, Clasp'd with thy dusky wing my aching head, While to imagination's startled ear Toll'd the slow bell, for bright Eliza dead.

"Stretched on her sable bier, the grave beside, A snow-white shroud her breathless bosom bound, O'er her wan brow the mimic lace was tied, And loves and virtues hung their garlands round.

"From those cold lips did softest accents flow? Round that pale mouth did sweetest dimples play? On this dull cheek the rose of beauty blow, And those dim eyes diffuse celestial day?

"Did this cold hand, unasking Want relieve, Or wake the lyre to every rapturous sound? How sad for other's woe this breast would heave! How light this heart for other's transport bound!

"Beats not the bell again?--Heavens, do I wake? Why heave my sighs, why gush my tears anew? Unreal forms my trembling doubts mistake, And frantic sorrow fears the vision true.

"Dreams to Eliza bend thy airy flight, Go, tell my

charmer all my tender fears, How love's fond woes alarm the silent night, And steep my pillow in unpitied tears."

Unwilling as I am to extend this memoir, I must give Miss Seward's criticism on the foregoing.

"The second verse of this charming elegy affords an instance of Dr. Darwin's too exclusive devotion to distinct picture in poetry; that it sometimes betrayed him into bringing objects so precisely to the eye as to lose in such precision their power of striking forcibly on the heart. The pathos in the second verse is much injured by the words 'mimic lace,' which allude to the perforated borders on the shroud. The expression is too minute for the solemnity of the subject. Certainly it cannot be natural for a shocked and agitated mind to observe, or to describe with such petty accuracy. Besides, the allusion is not sufficiently obvious. The reader pauses to consider what the poet means by 'mimic lace.' Such pauses deaden sensation and break the course of attention. A friend of the doctor's pleaded greatly that the line might run thus:--

"On her wan brow the *shadowy crape* was tied;"

but the alteration was rejected. Inattention to the rules of grammar in the first verse was also pointed out to him at the same time. The dream is addressed:

"Dread dream, that clasped my aching head,"

but nothing is said to it, and therefore the sense is

left unfinished, while the elegy proceeds to give a picture of the lifeless beauty. The same friend suggested a change which would have remedied the defect. Thus:--

"Dread *was the dream* that in the midnight air Clasped with its dusky wing my aching head, While to" &c., &c.

"Hence not only the grammatic error would have been done away, but the grating sound produced by the near alliteration of the harsh *dr* in '*dr*ead *dr*eam' removed, by placing those words at a greater distance from each other.

"This alteration was, for the same reason, rejected. The doctor would not spare the word *hovering*, which he said strengthened the picture; but surely the image ought not to be elaborately precise, by which a dream is transformed into an animal with black wings."[150]

Then Mrs. Pole got well, and the doctor wrote more verses and Miss Seward more criticism. It was not for nothing that Dr. Johnson came down to Lichfield.

In 1780 Colonel Pole died, and his widow, still young, handsome, witty, and--for those days--rich, was in no want of suitors.

"Colonel Pole," says Miss Seward, "had numbered twice the years of his fair wife. His temper was said to have been peevish and suspicious; yet not beneath those circumstances had her kind and cheerful attentions to him grown cold or remiss. He left her a jointure of 600*l.* per annum, a son to inherit his estate, and two female children amply portioned.

"Mrs. Pole, it has already been remarked, had much vivacity and sportive humour, with very engaging frankness of temper and manners. Early in her widowhood she was rallied in a large company upon Dr. Darwin's passion for her, and was asked what she would do with her captive philosopher. 'He is not very fond of churches, I believe,' said she, 'and even if he would go there for my sake, I shall scarcely follow him. He is too old for me.' 'Nay, Madam,' was the answer, 'what are fifteen years on the right side?' She replied, with an arch smile, 'I have had so *much* of that right side.'

"This confession was thought inauspicious for the doctor's hopes, but it did not prove so. The triumph of intellect was complete."[151]

Mrs. Pole had taken a strong dislike to Lichfield, and had made it a condition of her marriage that Dr. Darwin should not reside there after he had married her. In 1781, therefore, immediately after his marriage, he removed to Derby, and continued to live there till a fortnight before his death.

Here he wrote 'The Botanic Garden' and a great part of the 'Zoonomia.' Those who wish for a detailed analysis of 'The Botanic Garden' can hardly do better than turn to Miss Seward's pages. Opening them at random, I find the following:--

"The mention of Brindley, the father of commercial canals, has propriety as well as happiness. Similitude for their course to the sinuous track of a serpent, produces a fine picture of a gliding animal of that species, and it is succeeded by these supremely happy lines:--

"'So with strong arms immortal Brindley leads His long canals, and parts the velvet meads; Winding in lucid lines, the watery mass Mines the firm rock, or loads the deep morass;'[152] &c. &c. &c.

"The mechanism of the pump is next described with curious ingenuity. Common as is the machine, it is not unworthy a place in this splendid composition, as being, after the sinking of wells, the earliest of those inventions, which in situations of exterior aridness gave ready accession to water. This familiar object is illustrated by a picture of Maternal Beauty administering sustenance to her infant."[153]

Here we will leave the poetical part of the 'Botanic Garden.' The notes, however, to which are "still," as Dr. Dowson says, "instructive and amusing," and contain matter which, at the time they were written, was for the most part new.

Of the 'Zoonomia' there is no occasion to speak here, as a sufficient number of extracts from those parts that concern us as bearing upon evolution will be given presently.

On the 18th of April, 1802, Dr. Darwin had written "one page of a very sprightly letter to Mr. Edgeworth, describing the Priory and his purposed alterations there, when the fatal signal was given. He rang the bell and ordered the servant to send Mrs. Darwin to him. She came immediately, with his daughter, Miss Emma Darwin. They saw him shivering and pale. He desired them to send to Derby for his surgeon, Mr. Hadley. They did so, but all was over before he could arrive.

"It was reported at Lichfield that, perceiving himself growing rapidly worse, he said to Mrs. Darwin, 'My dear, you must bleed me instantly.' 'Alas! I dare not, lest--' 'Emma, will you? There is no time to be lost.' 'Yes, my dear father, if you will direct me.' At that moment he sank into his chair and expired."[154]

Dr. Dowson gives the letter to Mr. Edgeworth, which is as follows:--

"Dear Edgeworth,

"I am glad to find that you still amuse yourself with mechanism, in spite of the troubles of Ireland.

"The *use* of turning aside or downwards the claw of a table, I don't see; as it must then be reared against a wall, for it will not stand alone. If the use be for carriage, the feet may shut up, like the usual brass feet of a reflecting telescope.

"We have all been now removed from Derby about a fortnight, to the Priory, and all of us like our change of situation. We have a pleasant house, a good garden, ponds full of fish, and a pleasing valley, somewhat like Shenstone's--deep, umbrageous, and with a talkative stream running down it. Our house is near the top of the valley, well screened by hills from the east and north, and open to the south, where at four miles distance we see Derby tower.

"Four or more strong springs rise near the house, and have formed the valley which, like that of Petrarch, may be called Val Chiusa, as it begins, or is shut at the situation of the house. I hope you like the description, and hope farther that yourself and any part of your family will sometimes do us the pleasure of a visit.

"Pray tell the authoress" (Miss Maria Edgeworth) "that the water-nymphs of our valley will be happy to assist her next novel.

"My bookseller, Mr. Johnson, will not begin to print the 'Temple of Nature' till the price of paper is fixed by Parliament. I suppose the present duty is paid...."

At these words Dr. Darwin's pen stopped. What followed was written on the opposite side of the paper by another hand.

FOOTNOTES:

[137] 'Sketch, &c., of Erasmus Darwin,' pp. 3, 4.

[138] Miss Seward's 'Memoirs of Dr. Darwin,' p. 3.

[139] Ibid.

[140] Dr. Dowson's 'Sketch of Dr. Erasmus Darwin,' p. 50.

[141] Dr. Dowson's 'Sketch of Dr. Darwin,' p. 53.

[142] Miss Seward's 'Memoirs,' &c., p. 6.

[143] 'Memoirs,' &c., p. 14.

[144] 'Memoirs,' &c., p. 21.

[145] 'Memoirs,' &c., p. 62.

[146] 'Memoirs,' &c., p. 68.

[147] Miss Seward's 'Memoirs,' p. 69.

[148] 'Memoirs,' &c., p. 84.

[149] Ibid., p. 105.

[150] 'Memoirs,' &c., p. 120.

[151] 'Memoirs,' &c., p. 149.

[152] 'Memoirs,' &c., p. 249.

[153] 'Memoirs,' &c., p. 250.

[154] 'Memoirs,' &c., p. 426.

CHAPTER XIII.

PHILOSOPHY OF DR. ERASMUS DARWIN.

Considering the wide reputation enjoyed by Dr. Darwin at the beginning of this century, it is surprising how completely he has been lost sight of. The 'Botanic Garden' was translated into Portuguese in 1803; the 'Loves of the Plants' into French and Italian in 1800 and 1805; while, as I have already said, the 'Zoonomia' had appeared some years earlier in Germany. Paley's 'Natural Theology' is written throughout at the 'Zoonomia,' though he is careful, *more suo*, never to mention this work by name. Paley's success was probably one of the chief causes of the neglect into which the Buffonian and Darwinian systems fell in this country. Dr. Darwin is as reticent about teleology as Buffon, and presumably for the same reason, but the evidence in favour of design was too obvious; Paley, therefore, with his usual keen-sightedness seized upon this weak point, and had the battle all his own way, for Dr. Darwin died the same year as that in which the 'Natural Theology' appeared. The unfortunate failure to see that evolution involves design and purpose as necessarily and far more intelligibly than the theological view of creation, has retarded our perception of many important facts for three-quarters of a century.

However this may be, Dr. Darwin's name has been but little before the public during the controversies of the last thirty years. Mr. Charles Darwin, indeed, in the "historical sketch" which he has prefixed to the later editions of his 'Origin of Species,' says, "It is curious how largely my grandfather, Dr. Erasmus Darwin, anticipated the views and erroneous grounds of opinion of Lamarck in his 'Zoonomia,' vol. i. pp. 500-510, published in 1794."[155] And a few lines lower Mr. Darwin adds, "It is rather a singular instance of the manner in which similar views arise at about the same time, that Goethe in Germany, and Geoffroy St. Hilaire (as we shall immediately see) in France, came to the same conclusion on the 'Origin of Species' in the years 1794-1796." Acquaintance with Buffon's work will explain much of the singularity, while those who have any knowledge of the writings of Dr. Darwin and Étienne Geoffroy St. Hilaire will be aware that neither would admit the other as "coming to the same conclusion," or even nearly so, as himself. Dr. Darwin goes beyond his successor, Lamarck, while Étienne Geoffroy does not even go so far as Dr. Darwin's predecessor, Buffon, had thought fit to let himself be known as going. I have found no other reference to Dr. Darwin in the 'Origin of Species,' except the two just given from the same note. In the first edition I find no mention of him.

The chief fault to be found with Dr. Darwin's treatise on evolution is that there is not enough of it; what there is, so far from being "erroneous," is admirable. But so great a subject should have had a book to itself, and not a mere fraction of a book. If his opponents, not venturing to dispute with him, passed over one book in silence, he should have followed it up with another, and another, and another, year by year, as Buffon and Lamarck did; it is only thus that men can expect to succeed against vested interests. Dr. Darwin could speak with a freedom that was denied to Buffon. He took Buffon at his word as well as he could, and carried out his principles to what he conceived to be their logical conclusion. This was doubtless what Buffon had desired and reckoned on, but, as I have said already, I question how far Dr. Darwin understood Buffon's humour; he does not present any of the phenomena of having done so, and therefore I am afraid he must be said to have missed it.

Like Buffon, Dr. Darwin had no wish to see far beyond the obvious; he missed good things sometimes, but he gained more than he lost; he knew that it is always on the margin, as it were, of the self-evident that the greatest purchase against the nearest difficulty is obtainable. His life was not one of Herculean effort, but, like the lives of all those organisms that are most likely to develop and transmit a useful modification, it was one of well-sustained activity; it was a long-continued keeping open of the windows of his own mind, much after the advice he gave to the Nottingham weavers. Dr. Darwin knew, and, I imagine, quite instinctively, that nothing tends to oversight like overseeing. He does not trouble himself about the origin of life; as for the perceptions and reasoning faculties of animals and plants, it is enough for him that animals and plants do things which we say involve sensation and consciousness when we do them ourselves or see others do them. If, then, plants and animals appear as if they felt and understood, let the matter rest there, and let us say they feel and understand--being guided by the common use of language, rather than by any theories concerning brain and nervous system. If any young writer happens to be in want of a subject, I beg to suggest that he may find his opportunity in a 'Philosophy of the Superficial.'

Though Dr. Darwin was more deeply impressed than Buffon with the oneness of personality between parents and offspring, so that these latter are not "new" creatures, but "elongations of the parents," and hence "may retain some of the habits of the parent system," he did not go on to infer definitely all that he might easily have inferred from such a pregnant premiss. He did not refer the repetition by offspring, of actions which their parents have done for many generations, but which they can never have seen those parents do, to the memory (in the strict sense of the word) of their having done those actions when they were in the persons of their parents; which memory, though dormant until awakened by the presence of associated ideas, becomes promptly kindled into activity when a sufficient number of these ideas are reproduced.

This, I gather, is the theory put forward by Professor Hering, of whose work, however, I know no more than is told us by Professor Ray Lankester in an article which, appeared in 'Nature,' July 13th, 1876. This theory seems to be adopted by Professor Haeckel, and to receive support from Professor Ray Lankester himself. Knowing no German, I have been unable to make myself acquainted with Professor Hering's position in detail, but its similarity to, if not identity with, that taken by myself subsequently, but independently, in 'Life and Habit,' seems sufficiently established by the following extracts; it is to be wished, however, that a full account of this lecture were accessible to English readers. The extracts are as follows:--

"Professor Hering has the merit of introducing some striking phraseology into his treatment of the subject which serves to emphasize the leading idea. He points out that since all transmission of 'qualities' from cell to cell in the growth and repair of one and the same organ, or from parent to offspring, is a transmission of vibrations or affections of material particles, whether these qualities manifest themselves as form, or as a facility for entering on a given series of vibrations, we may speak of all such phenomena as 'memory,' whether it be the conscious memory exhibited by the nerve cells of the brain or the unconscious memory we call habit, or the inherited memory we call instinct; or whether, again, it be the reproduction of parental form and minute structure. All equally may be called the 'memory of living matter.' From the earliest existence of protoplasm to the present day the memory of living matter is continuous. Though individuals die, the universal memory of living matter is carried on.

"Professor Hering, in short, helps us to a comprehensive conception of the nature of heredity and adaptation, by giving us the term 'memory' conscious or unconscious, for the continuity of Mr. Herbert Spencer's polar forces, or polarities of physiological units.

"The undulatory movement of the plastidules is the key to the mechanical explanation of all the essential phenomena of life. The plastidules are liable to have their undulations affected by every external force, and, once modified, the movement does not return to its pristine condition. By assimilation they continually increase to a certain point in size, and then divide, and thus perpetuate in the undulatory movement of successive generations, the impressions or resultants due to the action of external agencies on individual plastidules. This is Memory. All plastidules possess memory; and Memory which we see in its ultimate analysis is identical with reproduction, is the distinguishing feature of the plastidule; is that which it alone of all molecules possesses, in addition to the ordinary properties of the physicist's molecule; is, in fact, that which distinguishes it as vital. To the sensitiveness of the movement of plastidules is due Variability--to their unconscious Memory the power of Hereditary Transmission. As we know them to-day they may 'have learnt little, and forgotten nothing' in one organism, and 'have learnt much, and forgotten much' in another; but in all, their memory if sometimes fragmentary, yet reaches back to the dawn of life upon the earth.--E. Ray Lankester."

Nothing can well be plainer and more uncompromising than the above. Professor Hering would, I gather, no less than myself, refer the building of its nest by a bird to the intense--but unconscious, owing to its very perfection and intensity--recollection by the bird of the nests it built when it was in the persons of its ancestors; this memory would begin to stimulate action when the surrounding associations, such as temperature, state of vegetation, &c., reminded it of the time when it had been in the habit of beginning to build in countless past generations. Dr. Darwin does not go so far as this. He says that wild birds choose spring as their building time "from their *acquired* knowledge that the mild temperature of the air is more convenient for hatching their eggs," and a little lower down he speaks of the fact that graminivorous animals generally produce their young in spring, as "part of the traditional knowledge which they learn *from the example* of their parents."[156]

Again he says, that birds "seem to be instructed how to build their nests *from their observation* of that in which they were educated, and from their knowledge of those things that are most agreeable to their touch in respect to warmth, cleanliness, and stability."

Had Dr. Darwin laid firmly hold of two superficial facts concerning memory which we can all of us test for ourselves--I mean its dormancy until kindled by the return of a sufficient number of associated ideas, and its unselfconsciousness upon becoming intense and perfect--and had he connected these two facts with the unity of life through successive generations--an idea which plainly haunted him--he would have been saved from having to refer instinct to imitation, in the face of the fact that in a thousand instances the creature imitating can never have seen its model, save when it was a part of its parents,--seeing what they saw, doing what they did, feeling as they felt, and remembering what they remembered.

Miss Seward tells us that Dr. Darwin read his chapter on instinct "to a lady who was in the habit of rearing canary birds. She observed that the pair which he then saw building their nest in her cage, were a male and female, who had been hatched and reared in that very *cage*, and were not in existence when the mossy cradle was fabricated in which *they* first saw light." She asked him, and quite reasonably, "how, upon his principle of imitation, he could account for the nest he then saw building, being constructed even to the precise disposal of every hair and shred of wool upon the model of *that* in which the pair were born, and on which every other canary bird's nest is constructed, when the proper materials are furnished. That of the pyefinch," she added, "is of much compacter form, warmer, and more comfortable. Pull one of these nests to pieces for its materials; and place another nest before these canary birds as a pattern, and see if they will make the slightest attempt to imitate their model! No, the result of their labour will, upon instinctive hereditary impulse, be exactly the slovenly little mansion of their race, the same with that which their parents built before themselves were hatched. The Doctor could not do away the force of that single fact, with which his system was incompatible, yet he maintained that system with philosophic sturdiness, though experience brought confutation from a thousand sources."[157]

As commonly happens in such disputes, both were right and both were wrong. The lady was right in refusing to refer instinct to imitation, and the Doctor was right in maintaining reason and instinct to be but different degrees of perfection of the same mental processes. Had he substituted "memory" for "imitation," and asked the lady to define "sameness" or "personal identity," he would have soon secured his victory.

The main fact, compared with which all else is a matter of detail, is the admission that instinct is only reason become habitual. This admission involves, consciously or unconsciously, the admission of all the principles contended for in 'Life and Habit'; principles which, if admitted, make the facts of heredity intelligible by showing that they are of the same character as other facts which we call intelligible, but denial of which makes nonsense of half the terms in common use concerning it. For the view that instinct is habitual reason involves sameness of personality and memory as common to parents and offspring; it involves also the latency of that memory till rekindled by the return of a sufficient number of its associated ideas, and points the unconsciousness with which habitual actions are performed. These principles being grasped, the infertility *inter se* of widely distant species, the commonly observed sterility of hybrids, the sterility of certain animals and plants under confinement, the phenomena of old age as well as those of growth, and the principle which underlies longevity and alternate generations, follow logically and coherently, as I showed in 'Life and Habit.' Moreover, we find that the terms in common use show an unconscious sense that some such view as I have insisted on was wanted and would come, for we find them made and to hand already; few if any will require altering; all that is necessary is to take common words according to their common meanings.

Dr. Darwin is very good on this head. Here, as everywhere throughout his work, if things or qualities appear to resemble one another sufficiently and without such traits of unlikeness, on closer inspection, as shall destroy the likeness which was apparent at first, he connects them, all theories notwithstanding. I have given two instances of his manner of looking at instinct and reason.[158] "If these are not," he concludes, "deductions *from their own previous experience, or observation*, all the actions of mankind must be resolved into instincts."[159]

If by "previous experience" we could be sure that Dr. Darwin persistently meant "previous experience in the persons of their ancestors," he would be in an impregnable position. As it is, we feel that though he had caught sight of the truth, and had even held it in his hands, yet somehow or other it just managed to slip through his fingers.

Again he writes:--

"So flies burn themselves in candles, deceived like mankind by the misapplication of their knowledge."

Again:--

"An ingenious philosopher has lately denied that animals can enter into contracts, and thinks this an essential difference between them and the human creature: but does not daily observation convince us that they form contracts of friendship with each other and with mankind? When puppies and kittens play together is there not a tacit contract that they will not hurt each other? And does not your favourite dog expect you should give him his daily food for his services and attention to you? And thus barters his love for your protection? In the same manner that all contracts are made among men that do not understand each other's arbitrary language."[160]

One more extract from a chapter full of excellent passages must suffice.

"One circumstance I shall relate which fell under my own eye, and showed the power of reason in a wasp, as it is exercised among men. A wasp on a gravel walk had caught a fly nearly as large as himself; kneeling on the ground, I observed him separate the tail and the head from the body part, to which the wings were attached. He then took the body part in his paws, and rose about two feet from the ground with it; but a gentle breeze wafting the wings of the fly turned him round in the air, and he settled again with his prey upon the gravel. I then distinctly observed him cut off with his mouth first one of the wings and then the other, after which he flew away with it, unmolested by the wind.

"Go, proud reasoner, and call the worm thy sister!"[161]

Dr. Darwin's views on the essential unity of animal and vegetable life are put forward in the following admirable chapter on "Vegetable Animation," which I will give in full, and which is confirmed in all important respects by the latest conclusions of our best modern scientists, so, at least, I gather from Mr. Francis Darwin's interesting lecture.[162]

"I. 1. The fibres of the vegetable world, as well as those of the animal, are excitable into a variety of motion by irritations of external objects. This appears particularly in the mimosa or sensitive plant, whose leaves contract on the slightest injury: the *Dionæa muscipula*, which was lately brought over from the marshes of America, presents us with another curious instance of vegetable irritability; its leaves are armed with spines on their upper edge, and are spread on the ground around the stem; when an insect creeps on any of them in its passage to the flower or seed, the leaf shuts up like a steel rat-trap, and destroys its enemy.[163]

"The various secretions of vegetables as of odour, fruit, gum, resin, wax, honey, seem brought about in the same manner as in the glands of animals; the tasteless moisture of the earth is converted by the hop plant into a bitter juice; as by the caterpillar in the nutshell, the sweet powder is converted into a bitter powder. While the power of absorption in the roots and barks of vegetables is excited into action by the fluids applied to their mouths like the lacteals and lymphatics of animals.

"2. The individuals of the vegetable world may be considered as inferior or less perfect animals; a tree is a congeries of many living buds, and in this respect resembles the branches of the coralline, which are a congeries of a multitude of animals. Each of these buds of a tree has its proper leaves or petals for lungs, produces its viviparous or its oviparous offspring in buds or seeds; has its own roots, which, extending down the stem of the tree, are interwoven with the roots of the other buds, and form the bark, which is the only living part of the stem, is annually renewed and is superinduced upon the former bark, which then dies, and, with its stagnated juices gradually hardening into wood, forms the concentric circles which we see in blocks of timber.

"The following circumstances evince the individuality of the buds of trees. First, there are many trees whose whole internal wood is perished, and yet the branches are vegete and healthy. Secondly, the fibres of the bark of trees are chiefly longitudinal, resembling roots, as is beautifully seen in those prepared barks that were lately brought from Otaheita. Thirdly, in horizontal wounds of the bark of trees, the fibres of the upper lip are always elongated downwards like roots, but those of the lower lip do not approach to meet them. Fourthly, if you wrap wet moss round any joint of a vine, or cover it with moist earth, roots will shoot out from it. Fifthly, by the inoculation or engrafting of trees many fruits are produced from one stem. Sixthly, a new tree is produced from a branch plucked from an old one and set in the ground. Whence it appears that the buds of deciduous trees are so many annual plants, that the bark is a contexture of the roots of each individual bud, and that the internal wood is of no other use but to support them in the air, and that thus they resemble the animal world in their individuality.

"The irritability of plants, like that of animals, appears liable to be increased or decreased by habit; for those trees or shrubs which are brought from a colder climate to a warmer, put out their leaves and blossoms a fortnight sooner than the indigenous ones.

"Professor Kalm, in his travels in New York, observes that the apple trees brought from England blossom a fortnight sooner than the native ones. In our country, the shrubs that are brought a degree or two from the north are observed to flourish better than those which come from the south. The Siberian barley and cabbage are said to grow larger in this climate than the similar more southern vegetables; and our hoards of roots, as of potatoes and onions, germinate with less heat in spring, after they have been accustomed to the winter's cold, than in autumn, after the summer's heat.

"II. The stamens and pistils of flowers show evident marks of sensibility, not only from many of the stamens and some pistils approaching towards each other at the season of impregnation, but from many of them closing their petals and calyxes during the cold part of the day. For this cannot be ascribed to irritation, because cold means a defect of the stimulus of heat; but as the want of accustomed stimuli produces pain, as in coldness, hunger, and thirst of animals, these motions of vegetables in closing up their flowers must be ascribed to the disagreeable sensation, and not to the irritation of cold. Others close up their leaves during darkness, which, like the former, cannot be owing to irritation, as the irritating material is withdrawn.

"The approach of the anthers in many flowers to the stigmas, and of the pistils of some flowers to the anthers, must be ascribed to the passion of love, and hence belongs to sensation, not to irritation.

"III. That the vegetable world possesses some degree of voluntary powers appears from their necessity to sleep, which we have shown in Section XVIII. to consist in the temporary abolition of voluntary power. This voluntary power seems to be exerted in the circular movement of the tendrils of the vines, and other climbing vegetables; or in the efforts to turn the upper surfaces of their leaves, or their flowers, to the light.

"IV. The associations of fibrous motions are observable in the vegetable world as well as in the animal. The divisions of the leaves of the sensitive plant have been accustomed to contract at the same time from the absence of light; hence, if by any other circumstance, as a slight stroke or injury, one division is irritated into contraction, the neighbouring ones contract also from their motions being associated with those of the irritated part. So the various stamina of the class of syngenesia have been accustomed to contract together in the evening, and thence if you stimulate any one of them with a pin, according to the experiment of M. Colvolo, they all contract from their acquired associations.

"To evince that the collapsing of the sensitive plant is not owing to any mechanical vibrations propagated along the whole branch when a single leaf is struck with the finger, a leaf of it was slit with sharp scissors, with as little disturbance as possible, and some seconds of time passed before the plant seemed sensible of the injury, and then the whole branch collapsed as far as the principal stem. This experiment was repeated several times with the least possible impulse to the plant.

"V. 1. For the numerous circumstances in which vegetable buds are analogous to animals, the reader is referred to the additional notes at the end of 'Botanic Garden,' Part I. It is there shown that the roots of vegetables resemble the lacteal system of animals; the sap vessels in the early spring, before their leaves expand, are analogous to the placental vessels of the foetus; that the leaves of land plants resemble lungs, and those of aquatic plants the gills of fish; that there are other systems of vessels resembling the vena portarum of quadrupeds, or the aorta of fish; that the digestive power of vegetables is similar to that of animals converting the fluids which they absorb into sugar;[164] that their seeds resemble the eggs of animals, and their buds and bulbs their viviparous offspring; and lastly, that the anthers and stigmas are real animals attached to their parent tree like polypi or coral insects, but capable of spontaneous motion; that they are affected with the passion of love, and furnished with powers of reproducing their species, and are fed with honey like the moths and butterflies which plunder their nectaries.[165]

"The male flowers of Vallisneria approach still nearer to apparent animality, as they detach themselves from the parent plant, and float on the surface of the water to the female ones.[166] Other flowers of the classes of monoecia and dioecia, and polygamia discharge the fecundating farina, which, floating in the air, is carried to the stigma of the female flowers, and that at considerable distances. Can this be effected by any specific attraction? Or, like the diffusion of the odorous particles of flowers, is it left to the currents of the winds, and the accidental miscarriages of it counteracted by the quantity of its production?

"2. This leads us to a curious inquiry, whether vegetables have ideas of external things? As all our ideas are originally received by our senses, the question may be changed to whether vegetables possess any organs of sense? Certain it is that they possess a sense of heat and cold, another of moisture and dryness, and another of light and darkness, for they close their petals occasionally from the presence of cold, moisture, or darkness. And it has been already shown that these actions cannot be performed simply from irritation, because cold and darkness are negative quantities, and on that account sensation, or volition are implied, and in consequence a sensorium or union of their nerves. So when we go into the light we contract the iris; not from any stimulus of the light on the fine muscles of the iris, but from its motions being associated with the sensation of too much light upon the retina, which could not take place without a sensorium or centre of union of the nerves of the iris, with those of vision.[167]

"Besides these organs of sense, which distinguish cold, moisture, and darkness, the leaves of mimosa, and of dionæa, and of drosera, and the stamens of many flowers, as of the berbery, and the numerous class of syngenesia, are sensible to mechanic impact, that is, they possess a sense of touch, as well as a common sensorium, by the medium of which their muscles are excited into action. Lastly, in many flowers the anthers, when mature, approach the stigma, in others the female organ approaches to the male. In a plant of collinsonia, a branch of which is now before me, the two yellow stamens are about three-eighths of an inch high, and diverge from each other at an angle of about fifteen degrees, the purple style is half an inch high, and in some flowers is now applied to the stamen on the right hand, and in others to that of the left; and will, I suppose, change place to-morrow in those, where the anthers have not yet effused their powder.

"I ask by what means are the anthers in many flowers and stigmas in other flowers directed to find their paramours? How do either of them know that the other exists in their vicinity? Is this curious kind of storge produced by mechanic attraction, or by the sensation of love? The latter opinion is supported by the strongest analogy, because a reproduction of the species is the consequence; and then another organ of sense must be wanted to direct these vegetable amourettes to find each other, one probably analogous to our sense of smell, which in the animal world directs the new-born infant to its source of nourishment, and they may thus possess a faculty of perceiving as well as of producing odours.

"Thus, besides a kind of taste at the extremity of their roots, similar to that of the extremities of our lacteal vessels, for the purpose of selecting their proper food, and besides different kinds of irritability residing in the various glands, which separate honey, wax, resin, and other juices from their blood; vegetable life seems to possess an organ of sense to distinguish the variations of heat, another to distinguish the varying degrees of moisture, another of light, another of touch, and probably another analogous to our sense of smell. To these must be added the indubitable evidence of their passion of love, and I think we may truly conclude that they are furnished with a common sensorium for each bud, and that they must occasionally repeat those perceptions, either in their dreams or waking hours, and consequently possess ideas of so many of the properties of the external world, and of their own existence."[168]

FOOTNOTES:

[155] 'Origin of Species,' note on p. xiv.

[156] 'Zoonomia,' vol. i. p. 170.

[157] Miss Seward's 'Memoirs,' &c., p. 491.

[158] See p. 116 of this volume.

[159] 'Zoonomia,' vol. i. p. 184.

[160] 'Zoonomia,' p. 171.

[161] 'Zoonomia,' p. 187.

[162] 'Nature,' March 14 and 21, 1878.

[163] See 'Botanic Garden,' part ii., note on Silene.

[164] 'On the Digestive Powers of Plants.' See Mr. Francis Darwin's lecture, already referred to.

[165] See 'Botanic Garden, part i., add. note, p. xxxix.

[166] Ibid., part ii., art. "Vallisneria."

[167] See 'Botanic Garden,' part i. cant 3, l. 440.

[168] 'Zoonomia,' vol. i. p. 107.

CHAPTER XIV.

FULLER QUOTATIONS FROM THE 'ZOONOMIA.'

The following are the passages in the 'Zoonomia' which have the most important bearing on evolution:--

"The ingenious Dr. Hartley, in his work on man, and some other philosophers have been of opinion, that our immortal part acquires during this life certain habits of action or of sentiment which become for ever indissoluble, continuing after death in a future state of existence; and add that if these habits are of the malevolent kind, they must render their possessor miserable even in Heaven. I would apply this ingenious idea to the generation or production of the embryon or new animal, which partakes so much of the form and propensities of its parent.

"*Owing to the imperfection of language the offspring is termed a new animal, but is in truth a branch or elongation of the parent, since a part of the embryon-animal is, or was, a part of the parent, and therefore in strict language, cannot be said to be entirely new at the time of its production; and, therefore, it may retain some of the habits of the parent system.*

"At the earliest period of its existence the embryon would seem to consist of a living filament with certain capabilities of irritation, sensation, volition, and association, and also with some acquired habits or propensities peculiar to the parents; the former of these are in common with other animals; the latter seem to distinguish or produce the kind of animal, whether man or quadruped, with the similarity of feature or form to the parent."[169]

Going on to describe the gradual development of the embryo, Dr. Darwin continues:--

"As the want of this oxygenation of the blood is perpetual (as appears from the incessant necessity of breathing by lungs or gills), the vessels become extended by the efforts of pain or desire to seek this necessary object of oxygenation, and to remove the disagreeable sensations which this want occasions."[170]

"The lateral production of plants by wires, while each new plant is thus chained to its parent, and continues to put forth another and another as the wire creeps onward on the ground, is exactly resembled by the tape-worm or tænia, so often found in the bowels, stretching itself in a chain quite from the stomach to the rectum. Linnæus asserts 'that it grows old at one extremity, while it continues to generate younger ones at the other, proceeding *ad infinitum* like a sort of grass; the separate joints are called gourd worms, and propagate new joints like the parent without end, each joint being furnished with its proper mouth and organs of digestion.'"[171]

"Many ingenious philosophers have found so great difficulty in conceiving the manner of the reproduction of animals, that they have supposed all the numerous progeny to have existed in miniature in the animal originally created; and that these infinitely minute forms are only evolved or distended, as the embryon increases in the womb. This idea, besides its being unsupported by any analogy we are acquainted with, ascribes a greater tenuity to organized matter than we can readily admit; as these included embryons are supposed each of them to consist of the various and complicate parts of animal bodies, they must possess a much greater degree of minuteness than that which was ascribed to the devils which tempted St. Anthony, of whom 20,000 were said to have been able to dance a saraband on the point of the finest needle without incommoding one another."[172]

"I conceive the primordium or rudiment of the embryon as secreted from the blood of the parent to consist of a simple living filament as a muscular fibre; which I suppose to be an extremity of a nerve of locomotion, as a fibre of the retina is an extremity of a nerve of sensation; as, for instance, one of the fibrils which compose the mouth of an absorbent vessel. I suppose this living filament of whatever form it may be, whether sphere, cube, or cylinder, to be endued with the capability of being excited into action by certain kinds of stimulus. By the stimulus of the surrounding fluid in which it is received from the male it may bend into a ring, and thus form the beginning of a tube. Such moving filaments and such rings are described by those who have attended to microscopic animalculæ. This living ring may now embrace or absorb a nutritive particle of the fluid in which it swims; and by drawing it into its pores, or joining it by compression to its extremities, may increase its own length or crassitude, and by degrees the living ring may become a living tube.

"the teleological impulse"

"With this new organization, or accretion of parts, new kinds of irritability may commence; for so long as there was but one living organ it could only be supposed to possess irritability; since sensibility may be conceived to be an extension of the effect of irritability over the rest of the system. These new kinds of irritability and of sensibility in consequence of new organization appear from variety of facts in the more mature animals; thus ... the lungs must be previously formed before their exertions to obtain fresh air can exist; the throat, or oesophagus, must be formed previous to the sensation or appetites of hunger and thirst, one of which seems to reside at the upper end and the other at the lower end of that canal."[173]

It seems to me Dr. Darwin is wrong in supposing that the organ must have preceded the power to use it. The organ and its use--the desire to do and the power to do--have always gone hand in hand, the organism finding itself able to do more according as it advanced its desires, and desiring to do more simultaneously with any increase in power, so that neither appetency nor organism can claim precedence, but power and desire must be considered as Siamese twins begotten together, conceived together, born together, and inseparable always from each other. At the same time they are torn by mutual jealousy; each claims, with some vain show of reason, to have been the elder brother; each intrigues incessantly from the beginning to the end of time to prevent the other from outstripping him; each is in turn successful, but each is doomed to death with the extinction of the other.

"So inflamed tendons and membranes, and even bones, acquire new sensations; and the parts of mutilated animals, as of wounded snails and polypi and crabs, are reproduced; and at the same time acquire sensations adapted to their situation. Thus when the head of a snail is reproduced after decollation with a sharp razor, those curious telescopic eyes are also reproduced, and acquire their sensibility to light, as well as their adapted muscles for retraction on the approach of injury.

"With every change, therefore, of organic form or

addition of organic parts, I suppose a new kind of irritability or of sensibility to be produced; such varieties of irritability or of sensibility exist in our adult state in the glands; every one of which is furnished with an irritability or a taste or appetency, and a consequent mode of action peculiar to itself.

"In this manner I conceive the vessels of the jaws to produce those of the teeth; those of the fingers to produce the nails; those of the skin to produce the hair; in the same manner as afterwards, about the age of puberty, the beard and other great changes in the form of the body and disposition of the mind are produced in consequence of new developments; for, if the animal is deprived of these developments, those changes do not take place. These changes I believe to be formed not by elongation or distension of primeval stamina, but by apposition of parts; as the mature crab fish when deprived of a limb, in a certain space of time, has power to regenerate it; and the tadpole puts forth its feet after its long exclusion from the spawn, and the caterpillar in changing into a butterfly acquires a new form with new powers, new sensations, and new desires."[174]

"From hence I conclude that with the acquisition of new parts, new sensations and new desires, as well as new powers are produced; and this by accretion to the old ones and not by distension of them. And finally, that the most essential parts of the system, as the brain for the purpose of distributing the powers of life, and the placenta for the purpose of oxygenating the blood, and the additional absorbent vessels, for the purpose of acquiring aliment, are first formed by the irritations above mentioned, and by the pleasurable sensations attending those irritations, and by the exertions in consequence of painful sensations similar to those of hunger and suffocation. After these an apparatus of limbs for future uses, or for the purpose of moving the body in its present natant state, and of lungs for future respiration, and of *testes* for future reproduction, are formed by the irritations and sensations and consequent exertions of the parts previously existing, and to which the new parts are to be attached.[175]

"The embryon" must "be supposed to be a living filament, which acquires or makes new parts, with new irritabilities as it advances in its growth."[176]

"From this account of reproduction it appears that all animals have a similar origin, viz. a single living filament; and that the difference of their forms and qualities has arisen only from the different irritabilities and sensibilities, or voluntarities, or associabilities, of this original living filament, and perhaps in some degree from the different forms of the particles of the fluids by which it has at first been stimulated into activity."[177]

purpose

"All animals, therefore, I contend, have a similar cause of their organization, originating from a single living filament, endued with different kinds of irritabilities and sensibilities, or of animal appetencies, which exist in every gland, and in every moving organ of the body, and are as essential to living organism as chemical affinities are to certain combinations of inanimate matter.

"If I might be indulged to make a simile in a philosophical work, I should say that the animal appetencies are not only perhaps less numerous originally than the chemical affinities, but that, like these latter, they change with every fresh combination; thus vital air and azote, when combined, produce nitrous acid, which now acquires the property of dissolving silver; so that with every new additional part to the embryon, as of the throat or lungs, I suppose a new animal appetency to be produced."[178]

Here, again, it should be insisted on that neither can the "additional part" precede "the appetency," nor the appetency precede the additional part for long together--the two advance nearly *pari passu*; sometimes the power a little ahead of the desire, stimulates the desire to an activity it would not otherwise have known; as those who have more money than they once had, feel new wants which they would not have known if they had not obtained the power to gratify them; sometimes, on the other hand, the desire is a little more active than the power, and pulls the power up to itself by means of the effort made to gratify the desire--as those who want a little more of this or that than they have money to pay for, will try all manner of shifts to earn the additional money they want, unless it is so much in excess of their present means that they give up the endeavour as hopeless; but whichever gets ahead, immediately sets to work to pull the other level with it, the getting ahead either of power or desire being exclusively the work of external agencies, while the coming up level of the other is due to agencies that are incorporate with the organism itself. Thus an unusually abundant supply of food, due to causes entirely beyond the control of the individual, is an external agency; it will immediately set power a little ahead of desire. On this the individual will eat as much as it can--thus learning *pro tanto* to be able to eat more, and to want more under ordinary circumstances--and will also breed rapidly up to the balance of the abundance. This is the work of the agencies incorporate in the organism, and will bring desire level with power again. Famine, on the other hand, puts desire ahead of power, and the incorporate agencies must either bring power up by resource and invention, or must pull desire back by eating less, both as individuals, and as the race, that is to say, by breeding less freely; for breeding is an assimilation of outside matter so closely akin to feeding, that it is only the feeding of the race, as against that of the individual.

I do not think the reader will find any clearer manner of picturing to himself the development of organism than by keeping the normal growth of wealth continually in his mind. He will find few of the phenomena of organic development which have not their counterpart in the acquisition of wealth. Thus a too sudden acquisition, owing to accidental and external circumstances and due to no internal source of energy, will be commonly lost in the next few generations. So a sudden sport due to a lucky accident of soil will not generally be perpetuated if the offspring plant be restored to its normal soil. Again, if the advance in power carry power suddenly far beyond any past desire, or be far greater than any past-remembered advance of power beyond desire--then desire will not come up level easily, but only with difficulty and all manner of extravagance, such as is likely to destroy the power itself. Demand and Supply are also good illustrations.

But to return to Dr. Darwin.

"When we revolve in our minds," he writes, "first the great changes which we see naturally produced in animals after their nativity, as in the production of the butterfly with painted wings from the crawling caterpillar; or of the respiring frog from the subnatant tadpole; from the boy to the bearded man, from the infant girl to the woman,--in both which cases mutilation will prevent due development.

"Secondly, when we think over the great changes introduced into various animals by artificial or accidental cultivation, as in horses, which we have exercised for the different purposes of strength or swiftness, in carrying burthens or in running races, or in dogs which have been cultivated for strength and courage, as the bull-dog; or for acuteness of his sense of smell, as the hound or spaniel; or for the swiftness of his foot, as the greyhound; or for his swimming in the water or for drawing snow sledges, as the rough-haired dogs of the north; or, lastly, as a play dog for children, as the lapdog; with the changes of the forms of the cattle which have been domesticated from the greatest antiquity, as camels and sheep, which have undergone so total a transformation that we are now ignorant from what species of wild animal they had their origin. Add to these the great changes of shape and colour which we daily see produced in smaller animals from our domestication of them, as rabbits or pigeons, or from the difference of climates and even of seasons; thus the sheep of warm climates are covered with hair instead of wool; and the hares and partridges of the latitudes which are long buried in snow become white during the winter months; add to these the various changes produced in the forms of mankind by their early modes of exertion, or by the diseases occasioned by their habits of life, both of which become hereditary, and that through many generations. Those who labour at the anvil, the oar, or the loom, as well as those who carry sedan chairs or who have been educated to dance upon the rope, are distinguishable by the shape of their limbs; and the diseases occasioned by intoxication deform the countenance with leprous eruptions, or the body with tumid viscera, or the joints with knots and distortions.

"Thirdly, when we enumerate the great changes produced in the species of animals before their nativity, as, for example, when the offspring reproduces the effects produced upon the parent by accident or cultivation; or the changes produced by the mixture of species, as in mules; or the changes produced probably by the exuberance of nourishment supplied to the fetus, as in monstrous births with additional limbs; many of these enormities of shape are propagated and continued as a variety at least, if not as a new species of animal. I have seen a breed of cats with an additional claw on every foot; of poultry also with an additional claw, and with wings to their feet; and of others without rumps. Mr. Buffon mentions a breed of dogs without tails which are common at Rome and Naples--which he supposes to have been produced by a custom long established of cutting their tails close off. There are many kinds of pigeons admired for their peculiarities which are more or less thus produced and propagated.[179]

"When we consider all these changes of animal form and innumerable others which may be collected from the books of natural history, we cannot but be convinced that the fetus or embryon is formed by apposition of new parts, and not by the distention of a primordial nest of germs included one within another like the cups of a conjurer.

"Fourthly, when we revolve in our minds the great similarity of structure which obtains in all the warm-blooded animals, as well quadrupeds, birds, and amphibious animals, as in mankind; from the mouse and bat to the elephant and whale; one is led to conclude that they have alike been produced from a similar living filament. In some this filament in its advance to maturity has acquired hands and fingers with a fine sense of touch, as in mankind. In others it has acquired claws or talons, as in tigers and eagles. In others, toes with an intervening web or membrane, as in seals and geese. In others it has acquired cloven hoofs, as in cows and swine; and whole hoofs in others, as in the horse: while in the bird kind this original living filament has put forth wings instead of arms or legs, and feathers instead of hair. In some it has protruded horns on the forehead instead of teeth in the fore part of the upper jaw; in others, tusks instead of horns; and in the others, beaks instead of either. And all this exactly as is seen daily in the transmutation of the tadpole, which acquires legs and lungs when he wants them, and loses his tail when it is no longer of service to him.

"Fifthly, from their first rudiment or primordium to the termination of their lives, all animals undergo perpetual transformations; *which are in part produced by their own exertions in consequence of their desires and aversions, of their pleasures and their pains, or of irritations or of associations; and many of these acquired forms or propensities are transmitted to their posterity.*

"As air and water are supplied to animals in sufficient profusion, the three great objects of desire which have changed the forms of many animals by their desires to gratify them are those of lust, hunger, and security. A great want of one part of the animal world has consisted in the desire of the exclusive possession of the females; and these have acquired weapons to combat each other for this purpose, as the very thick, shield-like, horny skin on the shoulder of the boar is a defence only against animals of his own species who strike obliquely upwards, nor are his tusks for other purposes except to defend himself, as he is not naturally a carnivorous animal. So the horns of the stag are sharp to offend his adversary, but are branched for the purpose of parrying or receiving the thrust of horns similar to his own, and have therefore been formed for the purpose of combating other stags, for the exclusive possession of the females; who are observed like the ladies in the times of chivalry to attend the car of the victor.

"The birds which do not carry food to their young, and do not therefore marry, are armed with spurs for the purpose of fighting for the exclusive possession of the females, as cocks and quails. It is certain that these weapons are not provided for their defence against other adversaries, because the females of these species are without this armour. The final cause of this contest among the males seems to be *that the strongest and most active animal should propagate the species, which should thence become improved*."[180]

Dr. Darwin would have been on stronger ground if he had said that the *effect* of the contest among the males was that the fittest should survive, and hence transmit any fit modifications which had occurred to them as vitally true, rather than that the desire to attain this end had caused the contest; but either way the sentence just given is sufficient to show that he was not blind to the fact that the fittest commonly survive, and to the consequences of this fact. The use, however, of the word "thence," as well as of the expression "final cause," is loose, as Dr. Darwin would no doubt readily have admitted. Improvement in the species is due quite as much, by Dr. Darwin's own showing, to the causes which have led to such and such an animal's making itself the fittest, as to the fact that if fittest it will be more likely to survive and transmit its improvement. There have been two factors in modification; the one provides variations, the other accumulates them; neither can claim exclusive right to the word "thence," as though the modification was due to it and to it only. Dr. Darwin's use of the word "thence" here is clearly a slip, and nothing else; but it is one which brings him for the moment into the very error into which his grandson has fallen more disastrously.

"Another great want," he continues, "consists in the means of procuring food, which has diversified the forms of all species of animals. Thus the nose of the swine has become hard for the purpose of turning up the soil in search of insects and of roots. The trunk of the elephant is an elongation of the nose for the purpose of pulling down the branches of trees for his food, and for taking up water without bending his knees. Beasts of prey have acquired strong jaws or talons. Cattle have acquired a rough tongue and a rough palate to pull off the blades of grass, as cows and sheep. Some birds have acquired harder beaks to crack nuts, as the parrot. Others have acquired beaks to break the harder seeds, as sparrows. Others for the softer kinds of flowers, or the buds of trees, as the finches. Other birds have acquired long beaks to penetrate the moister soils in search of insects or roots, as woodcocks, and others broad ones to filtrate the water of lakes and to retain aquatic insects. All which seem to have been gradually produced during many generations *by the perpetual endeavour of the creature to supply the want of food, and to have been delivered to their posterity with constant improvement of them for the purposes required.*

"The third great want among animals is that of security, which seems to have diversified the forms of their bodies and the colour of them; these consist in the means of escaping other animals more powerful than themselves. Hence some animals have acquired wings instead of legs, as the smaller birds, for purposes of escape. Others, great length of fin or of membrane, as the flying fish and the bat. Others have acquired hard or armed shells, as the tortoise and the *Echinus marinus.*

"Mr. Osbeck, a pupil of Linnæus, mentions the American frog-fish, *Lophius Histrio*, which inhabits the large floating islands of sea-weed about the Cape of Good Hope, and has fulcra resembling leaves, that the fishes of prey may mistake it for the sea-weed, which it inhabits.[181]

"The contrivances for the purposes of security extend even to vegetables, as is seen in the wonderful and various means of their concealing or defending their honey from insects and their seeds from birds. On the other hand, swiftness of wing has been acquired by hawks and swallows to pursue their prey; and a proboscis of admirable structure has been acquired by the bee, the moth, and the humming bird for the purpose of plundering the nectaries of flowers. *All which seem to have been formed by the original living filament, excited into action by the necessities of the creatures which possess them*, and on which their existence depends.

"From thus meditating on the great similarity of the structure of the warm-blooded animals, and at the same time of the great changes they undergo both before and after their nativity; and by considering in how minute a portion of time many of the changes of animals above described have been produced;

would it be too bold to imagine that in the great length of time since the earth began to exist, perhaps millions of ages before the commencement of the history of mankind--would it be too bold to imagine that all warm-blooded animals have arisen from one living filament, which the Great First Cause endued with animality, with the power of attaining new parts, attended with new propensities, directed by irritations, sensations, volitions, and associations; and thus possessing the faculty of continuing to improve, by its own inherent activity, and of delivering down those improvements by generation to its posterity world without end!

"Sixthly, the cold-blooded animals, as the fish tribes, which are furnished with but one ventricle of the heart, and with gills instead of lungs, and with fins instead of feet or wings, bear a great similarity to each other; but they differ nevertheless so much in their general structure from the warm-blooded animals, that it may not seem probable at first view that the same living filament could have given origin to this kingdom of animals, as to the former. Yet are there some creatures which unite or partake of both these orders of animation, as the whales and seals; and more particularly the frog, who changes from an aquatic animal furnished with gills to an aerial one furnished with lungs.

"The numerous tribes of insects without wings, from the spider to the scorpion, from the flea to the lobster; or with wings, from the gnat or the ant to the wasp and the dragon-fly, differ so totally from each other, and from the red-blooded classes above described, both in the forms of their bodies and in their modes of life; besides the organ of sense, which they seem to possess in their antennæ or horns, to which it has been thought by some naturalists that other creatures have nothing similar; that it can scarcely be supposed that this nature of animals could have been produced by the same kind of living filament as the red-blooded classes above mentioned. And yet the changes which many of them undergo in their early state to that of their maturity, are as different as one animal can be from another. As those of the gnat, which passes his early state in water, and then stretching out his new wings and expanding his new lungs, rises in the air; as of the caterpillar and bee-nymph, which feed on vegetable leaves or farina, and at length bursting from their self-formed graves, become beautiful winged inhabitants of the skies, journeying from flower to flower, and nourished by the ambrosial food of honey.

"There is still another class of animals which are termed vermes by Linnæus, which are without feet or brain, and are hermaphrodites, as worms, leeches, snails, shell-fish, coralline insects, and sponges, which possess the simplest structure of all animals, and appear totally different from those already described. The simplicity of their structure, however, can afford no argument against their having been produced from a single living filament, as above contended.

"Last of all, the various tribes of vegetables are to be enumerated amongst the inferior orders of animals. Of these the anthers and stigmas have already been shown to possess some organs of sense, to be nourished by honey, and to have the power of generation like insects, and have thence been announced amongst the animal kingdom in Section XIII.; and to these must be added the buds and bulbs, which constitute the viviparous offspring of vegetation. The former I suppose to be beholden to a single living filament for their seminal or amatorial procreation; and the latter to the same cause for their lateral or branching generation, which they possess in common with the polypus, tænia, and volvox, and the simplicity of which is an argument in favour of the similarity of its cause.

"Linnæus supposes, in the introduction to his natural orders, that very few vegetables were at first created, and that their numbers were increased by their intermarriages, and adds, 'Suaderet hæc Creatoris leges a simplicibus ad composita.' Many other changes appear to have arisen in them by their perpetual contest for light and air above ground, and for food or moisture beneath the soil. As noted in the 'Botanic Garden,' Part II., note on Cuscuta. Other changes of vegetables from climate or other causes are remarked in the note on Curcuma in the same work. From these one might be led to imagine that each plant at first consisted of a single bulb or flower to each root, as the gentianella and daisy, and that in the contest for air and light, new buds grew on the old decaying flower-stem, shooting down their elongated roots to the ground, and that in process of ages tall trees were thus formed, and an individual bulb became a swarm of vegetables. Other plants which in this contest for light and air were too slender to rise by their own strength, learned by degrees to adhere to their neighbours, either by putting forth roots like the ivy, or by tendrils like the vine, or by spiral contortions like the honeysuckle, or by growing upon them like the mistleto, and taking nourishment from their barks, or by only lodging or adhering on them and deriving nourishment from the air as tillandsia.

"Shall we then say that the vegetable living filament was originally different from that of each tribe of animals above described? And that the productive living filament of each of those tribes was different from the other? Or as the earth and ocean were probably peopled with vegetable productions long before the existence of animals; and many families of these animals, long before other families of them, shall we conjecture *that one and the same kind of living filament is and has been the cause of all organic life*?[182]

"The late Mr. David Hume in his posthumous works places the powers of generation much above those of our boasted reason, and adds, that reason can only make a machine, as a clock or a ship, but the power of generation makes the maker of the machine; and probably from having observed that the greatest part of the earth has been formed out of organic recrements, as the immense beds of limestone, chalk, marble, from the shells of fish; and the extensive provinces of clay, sandstone, ironstone, coals, from decomposed vegetables; all of which have been first produced by generation, or by the secretion of organic life; he concludes that the world itself might have been generated rather than created; that it might have been gradually produced from very small beginnings, increasing by the activity of its inherent principles, rather than by a sudden evolution of the whole by the Almighty fire. What a magnificent idea of the infinite power of the great Architect! The Cause of causes! Parent of parents! Ens entium!"[183]

FOOTNOTES:

[169] 'Zoonomia,' vol. i. p. 484.

[170] Ibid. p. 485.

[171] Ibid. p. 493.

[172] 'Zoonomia,' vol. i. p. 494.

[173] 'Zoonomia,' vol. i. p. 497.

[174] 'Zoonomia,' vol. i. p. 498.

[175] 'Zoonomia,' vol. i. p. 500.

[176] Ibid. p. 501.

[177] Ibid. p. 502.

[178] 'Zoonomia,' vol. i. p. 503.

[179] 'Zoonomia,' vol. i. p. 505.

[180] 'Zoonomia,' vol. i. p. 507.

[181] 'Voyage to China,' p. 113.

[182] 'Zoonomia,' vol. i. p. 511.

[183] 'Zoonomia,' vol. i. p. 513.

CHAPTER XV.

MEMOIR OF LAMARCK.

I take the following memoir of Lamarck entirely from the biographical sketch prefixed by M. Martins to his excellent edition of the 'Philosophie Zoologique.'[184] From this sketch I find that "Lamarck was born August 1, 1744, at Barenton, in Picardy, being the eleventh child of Pierre de Monet, squire of the place, a man of old family, but poor. His father intended him for the Church, the ordinary resource of younger sons at that time, and accordingly placed him under the care of the Jesuits at Amiens. But this was not his vocation: the annals of his family spoke all to him of military glory; his eldest brother had died in the breaches at the siege of Bergen-op-Zoom; two others were still serving in the army, and France was exhausting her energies in an unequal struggle. His father would not yield to his wishes, but on his death, in 1760, Lamarck was left free to take his own line, and made his way at once--upon a very bad horse--to the army of Germany, then encamped at Lippstadt in Westphalia.

"He was the bearer of a letter written by Madame de Lameth, one of his neighbours in the country, and recommending him to M. de Lastic, colonel of the regiment of Beaujolais. This gentleman, on seeing before him a lad of seventeen, whose somewhat stunted growth made him look still younger than he really was, sent the youth immediately to his own quarters. The next day a battle was immediately impending, and M. de Lastic, on passing his regiment in review, saw his protégé in the first rank of a company of grenadiers. The French army was under the orders of the Marshal de Broglie and of the Prince de Soubise; the allied troops were commanded by Ferdinand of Brunswick. The two French generals were beaten owing to their divided counsels, and Lamarck's company, almost annihilated by the enemy's fire, was forgotten in the confusion of the retreat. All the officers, commissioned and non-commissioned, were killed, and only fourteen men out of the whole company remained alive: the eldest proposed to retreat, but Lamarck, improvising himself as commander, declared that they ought not to retire without orders. Presently the colonel seeing that this company did not rally sent an orderly officer who made his way up to it by protected paths. Next day Lamarck was made an officer, and shortly afterwards lieutenant.

"Fortunately for science," continues M. Martins, "this brilliant *début* was not to decide his career. After peace had been signed he was sent into garrison at Toulon and Monaco, where an inflammation of the lymphatic ganglions of the neck necessitated an operation which left him deeply scarred for life.

"The vegetation in the neighbourhood of Toulon

and Monaco now arrested the young officer's attention. He had already derived some little knowledge of botany from the '*Traité des Plantes usuelles*' of Chomel. Having retired from the service, and having nothing beyond his modest pension of four hundred francs a year, he took a situation at Paris with a banker; but drawn irresistibly to the study of nature, he used to study from his attic window the forms and movements of clouds, and made himself familiar with the plants in the Jardin du Roi or in the public gardens. He began to feel that he was on his right path, and understood, as Voltaire said of Condorcet, that discoveries of permanent value could make him no less illustrious than military glory.

"Dissatisfied with the botanical systems of his time, in six months he wrote his '*Flore française*,' preceded by the '*Clé dichotomique*,' with the help of which it is easy even for a beginner to arrive with certainty at the name of the plant before him." Of this work, M. Martins tells us in a note, that the second edition, published by Candolle in 1815, is still the standard work on French plants.

"In 1778 Rousseau had brought botany into vogue. Women and men of fashion took to it. Buffon had the three volumes of '*Flore française*' printed at the royal press, and in the following year Lamarck entered the Academy of Sciences. Buffon being anxious that his son should travel, gave him Lamarck for his companion and tutor. He thus made a trip through Holland, Germany, and Hungary, and became acquainted with Gleditsch at Berlin, with Jacquin at Vienna, and with Murray at Gottingen.

"The '*Encyclopédie méthodique*,' begun by Diderot and D'Alembert, was not yet completed. For this work Lamarck wrote four volumes, describing all the then known plants whose names began with the letters from A to P. This great work was completed by Poiret, and comprises twelve volumes, which appeared between the years 1783 and 1817. A still more important work, also part of the Encyclopedia, and continually quoted by botanists, is the '*Illustration des Genres*.' In this work Lamarck describes two thousand *genera*, and illustrates them, according to the title-page, with nine hundred engravings. Only a botanist can form any idea of the research in collections, gardens, and books, which such a work must have involved. But Lamarck's activity was inexhaustible. Sonnerat returned from India in 1781 with a very large number of dried plants; no one except Lamarck thought it worth while to inspect them, and Sonnerat, charmed with his enthusiasm, gave him the whole magnificent collection.

"In spite, however, of his incessant toil, Lamarck's position continued to be most precarious. He lived by his pen, as a publisher's hack, and it was with difficulty that he obtained even the poorly paid post of keeper of the king's cabinet of dried plants. Like most other naturalists he had thus to contend with incessant difficulties during a period of fifteen years.

"At length fortune bettered his condition while changing the direction of his labours. France was now under the Convention; what Carnot had done for the army Lakanal undertook to do for the natural sciences. At his suggestion a museum of natural history was established. Professors had been found for all the chairs save that of Zoology; but in that time of enthusiasm, so different from the present, France could find men of war and men of science wherever and whenever she had need of them. Étienne Geoffroy St. Hilaire was twenty-one years old, and was engaged in the study of mineralogy under Haüy. Daubenton said to him, 'I will undertake the responsibility for your inexperience. I have a father's authority over you. Take this professorship, and let us one day say that you have made zoology a French science.' Geoffroy accepted, and undertook the higher animals. Lakanal knew that a single professor could not suffice for the task of arranging the collections of the entire animal kingdom, and as Geoffroy was to class the vertebrate animals only, there remained the invertebrata--that is to say, insects, molluscs, worms, zoophytes--in a word, what was then the chaos of the unknown. 'Lamarck,' says M. Michelet, 'accepted the unknown.' He had devoted some attention to the study of shells with Bruguières, but he had still everything to learn, or I should perhaps say rather, everything to create in that unexplored territory into which Linnæus had declined to enter, and into which he had thus introduced none of the order he had so well known how to establish among the higher animals.

"Lamarck began his course of lectures at the museum in 1794, after a year's preparation, and at once established that great division of animals into vertebrate and invertebrate, which science has ever since recognized.

"Dividing the vertebrate animals--as Linnæus had already divided them--into mammals, birds, reptiles, and fishes, he divided the invertebrates into molluscs, insects, worms, echinoderms, and polyps. In 1799 he separated the crustacea from the insects, with which they had been classed hitherto; in 1800 he established the arachnids as a class distinct from the insects; in 1802 that of the annelids, a subdivision of the worms, and that of the radiata as distinct from the polyps. Time has approved the wisdom of these divisions, founded all of them upon the organic type of the creatures themselves--that is to say, upon the rational method introduced into zoology by Cuvier, Lamarck, and Geoffroy St. Hilaire.

"This introduction being devoted only to Lamarck's labours as a naturalist, we will pass over certain works in which he treats of physics and chemistry. These attempts--errors of a powerful mind which thought itself able by the help of pure reason to establish truths which rest only upon experience--attempts, moreover, which were some of them but resuscitations of exploded theories, such as that of 'phlogistic'--had not even the honour of being refuted: they did not deserve to be so, and should be a warning to all those who would write upon a subject without the necessary practical knowledge.

"At the beginning of this century there was not yet any such science as geology. People observed but little, and in lieu of observation made theories to embrace the entire globe. Lamarck made his in 1802, and twenty-three years later the judicious Cuvier still yielded to the prevailing custom in publishing his 'Discoveries on the Earth's Revolutions.'

"Lamarck's merit was to have discovered that there had been no catastrophes, but that the gradual action of forces during thousands of ages accounted for the changes observable upon the face of the earth, better than any sudden and violent perturbations. 'Nature,' he writes, 'has no difficulty on the score of time; she has it always at command; it is with her a boundless space in which she has room for the greatest as for the smallest operations.'"

Here we must not forget Buffon's fine passage, "Nature's great workman is Time," &c. See page 103.

"Lamarck," continues M. Martins, "was the first to distinguish littoral from ocean fossils, but no one accepts his theory that oceans make their beds deeper owing to the action of the tides, and distribute themselves differently over the earth's surface without any change of level of the different parts of that surface.

"Settling down to a single branch of science, in consequence of his professorship, Lamarck now devoted himself to the twofold labour of lecturing and classifying the collections at the museum. In 1802 he published his 'Considerations on the Organization of Living Bodies'; in 1809 his '*Philosophie Zoologique*,' a development of the 'Considerations'; and from 1816 to 1822 his Natural History of the invertebrate animals, in seven volumes. This is his great work, and, being entirely a work of description and classification, was received with the unanimous approbation of the scientific world. His 'Fossil Shells of the Neighbourhood of Paris'--a work in which his profound knowledge of existing shells enabled him to class with certainty the remains of forms that had disappeared thousands of ages ago--met also with a favourable reception.

"Lamarck was fifty years old before he began to study zoology; and prolonged microscopic examinations first fatigued and at length enfeebled his eyesight. The clouds which obscured it gradually thickened, and he became quite blind. Married four times, the father of seven children, he saw his small patrimony and even his earlier savings swallowed up by one of those hazardous investments with which promoters impose on the credulity of the public. His small endowment as professor alone protected him from destitution. Men of science whom his reputation as a botanist and zoologist had attracted near him, wondered at the manner in which he was neglected.

"He passed the last ten years of his laborious life in darkness, tended only by the affectionate care of his two daughters. The eldest wrote from his dictation part of the sixth and seventh volumes of his work on the invertebrate animals. From the time her father became confined to his room his daughter never left the house; and when first she did so after his death, she was distressed by the fresh air to which she had been so long a stranger.

"Lamarck died December 18, 1829, at the age of eighty-five. Latreille and Blainville were his successors at the museum. The incredible activity of the first professor had so greatly increased the number of the known invertebrata that it was found necessary to endow two professors, where one had originally been sufficient.

"His two daughters were left penniless. In the year 1832 I myself saw Mlle. Cornélie de Lamarck earning a scanty pittance by fastening dried plants on to paper, in the museum of which her father had been a professor. Many a species named and described by him must have passed under her eyes and increased the bitterness of her regret."[185]

FOOTNOTES:

[184] Paris, 1873.

[185] Introduction Biographique to M. Martins' edition of the 'Phil. Zool.,' pp. ix-xx.

CHAPTER XVI.

GENERAL MISCONCEPTION CONCERNING LAMARCK--HIS PHILOSOPHICAL POSITION.

"If Cuvier," says M. Isidore Geoffroy St. Hilaire,[186] "is the modern successor of Linnæus, so is Lamarck of Buffon. But Cuvier does not go so far as Linnæus, and Lamarck goes much farther than Buffon. Lamarck, moreover, took his own line, and his conjectures are not only much bolder, or rather more hazardous, but they are profoundly different from Buffon's.

"It is well known that the vast labours of Lamarck were divided between botany and physical science in the eighteenth century, and between zoology and natural philosophy in the nineteenth; it is, however, less generally known that Lamarck was long a partisan of the immutability of species. It was not till 1801, when he was already old, that he freed himself from the ideas then generally prevailing. But Lamarck, having once made up his mind, never changed it; in his ripe age he exhibits all the ardour of youth in propagating and defending his new convictions.

"In the three years, 1801, 1802, 1803, he enounced them twice in his lectures, and three times in his writings.[187] He returns to the subject and states his views precisely in 1806,[188] and in 1809 he devotes a great part of his principal work, the 'Philosophie Zoologique,' to their demonstration.[189] Here he might have rested and have quietly awaited the judgment of his peers; but he is too much convinced; he believes the future of science to depend so much upon his doctrine that to his dying day he feels compelled to explain it further and insist upon it. When already over seventy years of age he enounces it again, and maintains it as firmly as ever in 1815, in his 'Histoire des Animaux sans Vertèbres,' and in 1820 in his 'Système des Connaissances Positives.'[190]

"This doctrine, so dearly cherished by its author, and the conception, exposition, and defence of which so laboriously occupied the second half of his scientific career, has been assuredly too much admired by some, who have forgotten that Lamarck had a precursor, and that that precursor was Buffon. It has, on the other hand, been too severely condemned by others who have involved it in its entirety in broad and sweeping condemnation. As if it were possible that so great labour on the part of so great a naturalist should have led him to 'a fantastic conclusion' only--to a 'flighty error,' and, as has been often said, though not written, to 'one absurdity the more.' Such was the language which Lamarck heard during his protracted old age, saddened alike by the weight of years and blindness; this was what people did not hesitate to utter over his grave yet barely closed, and what, indeed, they are still saying--commonly, too, without any knowledge of what Lamarck maintained, but merely repeating at second hand bad caricatures of his teaching.

"When will the time come when we may see Lamarck's theory discussed--and, I may as well at once say, refuted in some important points--with at any rate the respect due to one of the most illustrious masters of our science? And when will this theory, the hardihood of which has been greatly exaggerated, become freed from the interpretations and commentaries by the false light of which so many naturalists have formed their opinion concerning it? If its author is to be condemned, let it be, at any rate, not before he has been heard."[191]

It is not necessary for me to give the extracts from Lamarck which M. Isidore Geoffroy St. Hilaire quotes in order to show what he really maintained, inasmuch as they will be given at greater length in the following chapter; but I may perhaps say that I have not found M. Geoffroy refuting Lamarck in any essential point.

Professor Haeckel says that to Lamarck "will always belong the immortal glory of having for the first time worked out the theory of descent as an independent scientific theory of the first order, and as the philosophical foundation of the whole science of Biology."

"The 'Philosophie Zoologique,'" continues Professor Haeckel, "is the first connected exposition of the theory of descent carried out strictly into all its consequences; ... and with the exception of Darwin's work, which appeared exactly half a century later, we know of none which we could in this respect place by the side of the 'Philosophie Zoologique.' How far it was in advance of its time is perhaps best seen from the circumstance that it was not understood by most men, and for fifty years was not spoken of at all."[192]

This is an exaggeration, both as regards the originality of Lamarck's work and the reception it has met with. It is probably more accurate to say with M. Martins that Lamarck's theory has "never yet had the honour of being discussed seriously,"[193] not, at least, in connection with the name of its originators.

So completely has this been so that the author of the 'Vestiges of Creation,' even in the edition of 1860, in which he unreservedly acknowledges the adoption of Lamarck's views, not unfrequently speaks disparagingly of Lamarck himself, and never gives him his due meed of recognition. I am not, therefore, wholly displeased to find this author conceiving himself to have been treated by Mr. Charles Darwin with some of the injustice which he has himself inflicted on Lamarck.

In the 1859 edition of the 'Origin of Species,' and in a very prominent place, Mr. Darwin says:--"The author of the 'Vestiges of Creation' would I presume say, that after a certain number of unknown generations, some bird had given birth to a woodpecker, and some plant to a misseltoe, and that these had been produced perfect as we now see them."[194] This is the only allusion to the 'Vestiges' which I have found in the first edition of the 'Origin of Species.'

Those who have read the 1853 edition of the 'Vestiges' will not be surprised to find the author rejoining, in his edition of 1860, that it was to be regretted Mr. Darwin should have read the 'Vestiges' "nearly as much amiss as though, like its declared opponents, he had an interest in misunderstanding it." And a little lower he adds that Mr. Darwin's book in no essential respect contradicts the 'Vestiges'; "on the contrary, while adding to its explanations of nature, it expresses substantially the same general ideas."[195] It is right to say that the passage thus objected to is not to be found in later editions of the 'Origin of

Species,' while in the historical sketch we now read as follows:--"In my opinion it (the 'Vestiges of Creation') has done excellent service in this country by calling attention to the subject, removing prejudice, and in thus preparing the ground for the reception of analogous views."

Mr. Darwin, the main part of whose work on the 'Origin of Species' is taken up with supporting the theory of descent with modification (which frequently in the recapitulation chapter of the 'Origin of Species' he seems to treat as synonymous with natural selection), has fallen into the common error of thinking that Lamarck can be ignored or passed over in a couple of sentences. I only find Lamarck's name twice in the 1859 edition of the 'Origin,' once on p. 242, where Mr. Darwin writes: "I am surprised that no one has advanced this demonstrative case of neuter insects, against the well-known doctrine of Lamarck;" and again, p. 427, where Lamarck is stated to have been the first to call attention to the "very important distinction between real affinities and analogical or adaptive resemblances." How far from demonstrative is the particular case which in 1859 Mr. Darwin considered so fatal to "the well-known doctrine of Lamarck"--which should surely, one would have thought, include the doctrine of descent with modification, which Mr. Darwin is himself supporting--I have attempted to show in 'Life and Habit,' but had perhaps better recapitulate briefly here.

Mr. Darwin writes: "In the simpler case of neuter insects all of one caste, *which, as I believe, have been rendered different from the fertile males and females through natural selection....*"[196] He thus attributes the sterility and peculiar characteristics, we will say, of the common hive working bees--"neuter insects all of one caste"--to natural selection. Now, nothing is more certain than that these characteristics--sterility, a cavity in the thigh for collecting wax, a proboscis for gathering honey, &c.--are due to the treatment which the eggs laid by the queen bee receive after they have left her body. Take an egg and treat it in a certain way, and it becomes a working bee; treat the same egg in a certain other way, and it becomes a queen. If the bees are in danger of becoming queenless they take eggs which were in the way of being developed into working bees, and change their food and cells, whereon they develop into queens instead. How Mr. Darwin could attribute the neutralization of the working bees--an act which is obviously one of abortion committed by the body politic of the hive on a balance of considerations--to the action of what he calls "natural selection," and how, again, he could suppose that what he was advancing had any but a confirmatory bearing upon Lamarck's position, is incomprehensible, unless the passage in question be taken as a mere slip. That attention has been called to it is plain, for the words "the well-known doctrine of Lamarck" have been changed in later editions into "the well-known doctrine of inherited habit as advanced by Lamarck,"[197] but this correction, though some apparent improvement on the original text, does little indeed in comparison with what is wanted.

Mr. Darwin has since introduced a paragraph

concerning Lamarck into the "historical sketch," already more than once referred to in these pages. In this he summarises the theory which I am about to lay before the reader, by saying that Lamarck "upheld the doctrine that all species, including man, are descended from other species." If Lamarck had been alive he would probably have preferred to see Mr. Darwin write that he upheld "the doctrine of descent with modification as the explanation of all differentiations of structure and instinct." Mr. Darwin continues, that Lamarck "seems" to have been chiefly led to his conclusion on the gradual change of species, "by the difficulty of distinguishing species and varieties, by the almost perfect gradation of forms in certain groups, and by the analogy of domestic productions."

Lamarck would probably have said that though he did indeed turn--as Mr. Darwin has done, and as Buffon and Dr. Darwin had done before him--to animals and plants under domestication, in illustration and support of the theory of descent with modification; and that though he did also insist, as so many other writers have done, on the arbitrary and artificial nature of the distinction between species and varieties, he was mainly led to agree with Buffon and Dr. Darwin by a broad survey of the animal kingdom, with the details also of which few naturalists have ever been better acquainted.

"Great," says Mr. Darwin, "is the power of steady misrepresentation,"--and greatly indeed has the just fame of Lamarck been eclipsed in consequence; "but," as Mr. Darwin finely continues, "the history of science shows that fortunately this power does not long endure."[198]

That Lamarck anticipated it, was prepared to face it, and even felt that things were thus, after all, as they should be, will appear from the shrewd and pleasant passage which is to be found near the close of his preface:--

"So great is the power of preconceived opinion, especially when any personal interest is enlisted on the same side as itself, that though it is hard to deduce new truths from the study of nature, it is still harder to get them recognized by other people.

"These difficulties, however, are on the whole more beneficial than hurtful to the cause of science; for it is through them that a number of eccentric, though perhaps plausible speculations, perish in their infancy, and are never again heard of. Sometimes, indeed, valuable ideas are thus lost; but it is better that a truth, when once caught sight of, should have to struggle for a long time without meeting the attention it deserves, than that every outcome of a heated imagination should be readily received.

"The more I reflect upon the numerous causes which affect our judgments, the more convinced I am that, with the exception of such physical and moral facts as no one can now throw doubt upon, all else is matter of opinion and argument; and we know well that there is hardly an argument to be found anywhere, against which another argument cannot plausibly be adduced. Hence, though it is plain that the various opinions of men differ greatly in probability and in the weight which should be attached to them, it seems to me that we are wrong when we blame those who differ from us.

"Are we then to recognize no opinions as well founded but those which are generally received? Nay--experience teaches us plainly that the highest and most cultivated minds must be at all times in an exceedingly small minority. No one can dispute this. Authority should be told by weight and not by number--but in good truth authority is a hard thing to weigh.

"Nor again--in spite of the many and severe conditions which a judgment must fulfil before it can be declared good--is it quite certain that those whom public opinion has declared to be authorities, are always right in the conclusions they arrive at.

"Positive facts are the only solid ground for man; the deductions he draws from them are a very different matter. Outside the facts of nature all is a question of probabilities, and the most that can be said is that some conclusions are more probable than others."

Lamarck's poverty was perhaps one main reason of the ease with which it was found possible to neglect his philosophical opinions. Science is not a kingdom into which a poor man can enter easily, if he happens to differ from a philosopher who gives good dinners, and has "his sisters and his cousins and his aunts" to play the part of chorus to him. Lamarck's two daughters do not appear to have been the kind of persons who could make effective sisters or cousins or aunts. Men of science are of like passions even with the other holy ones who have set themselves up in all ages as the pastors and prophets of mankind. The saint has commonly deemed it to be for the interests of saintliness that he should strain a point or two in his own favour--and the more so according as his reputation for an appearance of candour has been the better earned. If, then, Lamarck's opponents could keep choruses, while Lamarck had nothing to fall back upon but the merits of his case only, it is not surprising that he should have found himself neglected by the scientists of his own time. Moreover he was too old to have undertaken such an unequal contest. If he had been twenty years younger when he began it, he would probably have enjoyed his full measure of success before he died.

Not that Lamarck can claim, as a thinker, to stand on the same level with Dr. Darwin, and still less so with Buffon. He attempted to go too fast and too far. Seeing that if we accept descent with modification, the question arises whether what we call life and consciousness may not themselves be evolved from some thing or things which looked at one time so little living and conscious that we call them inanimate--and being anxious to see his theory reach, and to follow it, as far back as possible, he speculates about the origin of life; having formed a theory thereon, he is more inclined to interpret the phenomena of lower animal life so as to make them fit in with his theory, than as he would have interpreted them if there had been no theory at stake.

Thus his denial that sensation, and much more, intelligence and deliberate action, can exist without a brain and a nervous system, has led him to deny sensation, consciousness, and intelligence to many animals which act in such manner as would certainly have made him say that they feel and know what they are about, if he had formed no theory about brains and nervous systems.

Nothing can be more different than the manners in which Lamarck and Dr. Darwin wrote on this head. Lamarck over and over again maintains that where there is no nervous system there can be no sensation. Combating, for example, the assertion of Cabanis, that to live is to feel, he says that "the greater number of the polypi and all the infusoria, having no nervous system, it must be said of them as also of worms, that to live is still not to feel; and so again of plants."[199]

How different from this is the un-theory-ridden language of Dr. Darwin, quoted on p. 116 of this work.

Lamarck again writes:--

"The very imperfect animals of the lowest classes, having no nervous system, are simply irritable, have nothing but certain habits, experience no sensations, and never conceive ideas."

This, in the face of the performances of the amoeba--a minute jelly speck, without any special organ whatever--in making its tests, cannot be admitted. Is it possible that Lamarck was in some measure misled by believing Buffon to be in earnest when he advanced propositions little less monstrous?

"But," continues Lamarck, "the less imperfect animals which have a nervous system, though they have not the organ of intelligence, have instinct, habits, and proclivities; they feel sensations, and yet form no ideas whatever. I venture to say that where there is no organ for a faculty that faculty cannot exist."[200]

Who can tell what ideas a worm does or does not form? We can watch its actions, and see that they are such as involve what we call design and a perception of its own interest. Under these circumstances it seems better to call the worm a reasonable creature with Dr. Darwin than to say with Lamarck that because worms do not appear to have that organ which he assumes to be the sole means of causing sensation and ideas, therefore they can neither feel nor think. Doubtless they cannot feel and think as many sensations and thoughts as we can, but our ideas of what they can and cannot feel must be formed through consideration of what we see them do, and must be biassed by no theories of what they ought to be able to feel or not feel.

Again Lamarck, shortly after an excellent passage in which he points out that the lower animals gain by experience just as man does (and here probably he had in his mind the passage of Buffon referred to at p. 112 of this work), nevertheless writes:--

"If the facts and considerations put forward in this

volume be held worthy of attention, it will follow necessarily that there are some animals which have neither reason nor instinct" (I should be glad to see one of these animals and to watch its movements), "such as those which have no power of feeling; that there are others which have instinct but no degree whatever of reason" (whereas from Dr. Darwin's premises it should follow, and would doubtless be readily admitted by him, that instinct is reason, but reason many times repeated made perfect, and finally repeated by rote; so that far from being prior to reason, as Lamarck here implies, it can only come long afterwards), "such as those which have a system enabling them to feel, but which still lack the organ of intelligence; and finally, that there are those which have not only instinct, but over and above this a certain degree of reasoning power, such as those creatures which have one system for sensations and another for acts involving intelligence. Instinct is with these last animals the motive power of almost all their actions, and they rarely use what little reason they have. Man, who comes next above them, is also possessed of instincts which inspire some of his actions, but he can acquire much reason, and can use it so as to direct the greater part of his actions."[201]

All this will be felt to be less satisfactory than the simple directness of Dr. Darwin. It comes in great measure from following Buffon without being *en rapport* with him. On the other hand, Lamarck must be admitted to have elaborated the theory of "descent with modification" with no less clearness than Dr. Darwin, and with much greater fulness of detail. There is no substantial difference between the points they wish to establish; Dr. Darwin has the advantage in that not content with maintaining that there will be a power of adaptation to the conditions of an animal's existence which will determine its organism, he goes on to say what the principal conditions are, and shows more lucidly than Lamarck has done (though Lamarck adopts the same three causes in a passage which will follow), that struggle, and consequently modification, will be chiefly conversant about the means of subsistence, of reproduction, and of self-protection. Nevertheless, though Dr. Darwin has said enough to show that he had the whole thing clearly before him, and could have elaborated it as finely as or better than Lamarck himself has done, if he had been so minded, yet the palm must be given to Lamarck on the score of what he actually did, and this I observe to be the verdict of history, for whereas Lamarck's name is still daily quoted, Dr. Darwin's is seldom mentioned, and never with the applause which it deserves.

The resemblance between the two writers--that is to say, the complete coincidence of their views--is so remarkable that the question is forced upon us how far Lamarck knew the substance of Dr. Darwin's theory. Lamarck knew Buffon personally; he had been tutor to Buffon's son, and Buffon had three of Lamarck's volumes on the French Flora printed at the royal printing press;--how can we account for Lamarck's having had Buffon's theory of descent with modification before him for so many years, and yet remaining a partisan of immutability till 1801? Before this year we find no trace of his having accepted evolution; thenceforward he is one of the most ardent and constant exponents which this doctrine has ever had. What was it that repelled him in Buffon's system? How is it that in the 'Philosophie Zoologique' there is not, so far as I can remember, a single reference to Buffon, from whom, however, as we shall see, many paragraphs are taken with but very little alteration?

I am inclined to think that the secret of this sudden conversion must be found in a French translation by M. Deleuze of Dr. Darwin's poem, 'The Loves of the Plants' which appeared in 1800. Lamarck--the most eminent botanist of his time--was sure to have heard of and seen this, and would probably know the translator, who would be able to give him a fair idea of the 'Zoonomia.'

I will give a few of the passages which Lamarck would find in this translation. Speaking of Dr. Darwin, M. Deleuze says:--"Il falloit encore qu'un nouvel observateur, entrant dans la route qui venoit de s'ouvrir, s'y frayât des sentiers ignorés; que liant la physique végétale à la botanique il nous montrât dans les plantes, non seulement des corps organisés soumis à des lois constantes, mais des êtres doués sinon de sensibilité, au moins d'une irritabilité particulière, d'un principe de vie *qui leur fait exécuter des mouvements analogues à leurs besoins*....[202]

"Il est des animaux et des plantes qui par le laps du tems paroissent avoir éprouvé des changemens dans leur organisation, *pour s'accommoder à de nouveaux genres de nourriture et aux moyens de se la procurer*. Peut-être les productions de la nature font elles des progrès vers la perfection. Cette idée appuyée par les observations modernes sur l'accroissement progressif des parties solides du globe, s'accorde avec la dignité et la providence du créateur de l'univers."[203]

"La nature semble s'être fait un jeu d'établir entre tous les êtres organisés une sorte de guerre qui entretient leur activité: si elle a donné aux uns des moyens de défense, elle a donné aux autres des moyens d'attaque."[204]

Turning to the 'Botanic Garden' itself, I find that this admirable sentence belongs to M. Deleuze, and not to Dr. Darwin, who, however, has said what comes to much the same thing,[205] as may be seen p. 227 of this volume. But the authorship is immaterial; whether the passage was by Dr. Darwin or M. Deleuze, it was, in all probability, known to Lamarck before his change of front.

The note on Trapa Natans again[206] suggests itself as the source from which the passage in the 'Philosophie Zoologique' about the Ranunculus aquatilis is taken,[207] while one of the most important passages in the work, a summary, in fact, of the principal means of modification, seems to be taken, the first half of it from Buffon, and the second from Dr. Darwin. I have called attention to it on pp. 300, 301.

We may then suppose that Lamarck failed to understand Buffon, and conceived that he ought either to have gone much farther, or not so far; not being yet prepared to go the whole length himself, he opposed mutability till Dr. Darwin's additions to Buffon's ostensible theory reached him, whereon he at once adopted them, and having received nothing but a few notes and hints, felt himself at liberty to work the theory out independently and claim it. In so original a work as the '*Philosophie Zoologique*' must always be considered, this may be legitimate, but I find in it, as Isidore Geoffroy seems also to have found, a little more claim to complete independence than is acceptable to one who is fresh from Buffon and Dr. Darwin.

FOOTNOTES:

[186] 'Hist. Nat. Gén.,' tom. ii. p. 404, 1859.

[187] 'Système des Animaux sans Vertèbres,' Paris, in-8, an. ix. (1801); 'Discours d'Ouverture,' p. 12, &c.; 'Recherches sur l'Organisation des Corps Vivants,' Paris, in-8, 1802, p. 50, &c.; 'Discours d'Ouverture d'un Cours de Zoologie pour l'an ix.,' Paris, in-8, 1803. This discourse is entirely devoted to the consideration of the question, "What is Species?"

[188] 'Discours d'Ouverture d'un Cours de Zoologie,' 1806, Paris, in-8, p. 8, &c.

[189] See following chapter.

[190] 'Hist, des Anim. sans Vertèb.,' tom, i., Introduction, 1re ed., 1815; 'Syst. des Conn. Positives,' Paris, in-8, 1820, 1re part, 2me sect. ch. ii. p. 114, &c.

[191] 'Hist. Nat. Gén.,' tom. ii. p. 407.

[192] 'History of Creation,' English translation, vol. i. pp. 111, 112.

[193] M. Martins' edition of the 'Philosophie Zoologique,' Paris, 1873. Introd., p. vi.

[194] 'Origin of Species,' p. 3, 1859.

[195] 'Vestiges of Creation,' ed. 1860, Proofs, Illustrations, &c., p. lxiv.

[196] 'Origin of Species,' ed. 1, p. 239; ed. 6, p. 231.

[197] 'Origin of Species,' ed. 1, p. 242; ed. 6, 1876, p. 233.

[198] 'Origin of Species,' p. 421, ed. 1876.

[199] 'Phil. Zool.,' vol. i. p. 404.

[200] Ibid. vol. ii. p. 324.

[201] 'Phil. Zool.,' vol. ii. p. 410.

[202] 'Les Amours des Plantes,' Discours Prélim., p. 7. Paris, 1800.

[203] Ibid., Notes du chant i., p. 202.

[204] Ibid. p. 238.

[205] 'Zoonomia,' vol. i. p. 507.

[206] 'Les Amours des Plantes,' p. 360.

[207] Vol. i. p. 231, ed. M. Martins, 1873.

CHAPTER XVII.

SUMMARY OF THE 'PHILOSOPHIE ZOOLOGIQUE.'

The first part of the '*Philosophie Zoologique*' is the one which deals with the doctrine of evolution or descent with modification. It is to this, therefore, that our attention will be confined. Yet only a comparatively small part of the three hundred and fifty pages which constitute Lamarck's first part are devoted to setting forth the reasons which led him to arrive at his conclusions--the greater part of the volume being occupied with the classification of animals, which we may again omit, as foreign to our purpose.

I shall condense whenever I can, but I do not think the reader will find that I have left out much that bears upon the argument. I shall also use inverted commas while translating with such freedom as to omit several lines together, where I can do so without suppressing anything essential to the elucidation of Lamarck's meaning. I shall, however, throughout refer the reader to the page of the original work from which I am translating.

"The common origin of bodily and mental

phenomena," says Lamarck in his preliminary chapter, "has been obscured, because we have studied them chiefly in man, who, as the most highly developed of living beings, presents the problem in its most difficult and complicated aspect. If we had begun our study with that of the lowest organisms, and had proceeded from these to the more complex ones, we should have seen the progression which is observable in organization, and the successive acquisition of various special organs, with new faculties for every additional organ. We should thus have seen that sense of needs--originally hardly perceptible, but gradually increasing in intensity and variety--has led to the attempt to gratify them; that the actions thus induced, having become habitual and energetic, have occasioned the development of organs adapted for their performance; that the force which excites organic movements can in the case of the lowest animals exist outside them and yet animate them; that this force was subsequently introduced into the animals themselves, and fixed within them; and, lastly, that it gave rise to sensibility and, in the end, to intelligence."[208] The reader had better be on his guard here, and whenever Lamarck is speculating about the lowest forms of action and sensation. I have thought it well, however, to give enough of these speculations, as occasion arises, to show their tendency.

"Sensation is not the proximate cause of organic movements. It may be so with the higher animals, but it cannot be shown to be so with plants, nor even with all known animals. At the outset of life there was none of that sensation which could only arise where organic beings had already attained a considerable development. Nature has done all by slow gradations, both organs and faculties being the outcome of a progressive development.[209]

"The mere composition of an animal is but a small part of what deserves study in connection with the animal itself. The effects of its surroundings in causing new wants, the effects of its wants in giving rise to actions, those of its actions in developing habits and tendencies, the effects of use and disuse as affecting any organ, the means which nature takes to preserve and make perfect what has been already acquired--these are all matters of the highest importance.[210]

"In their bearing upon these questions the invertebrate animals are more important and interesting than the vertebrate, for they are more in number, and being more in number are more varied; their variations are more marked, and the steps by which they have advanced in complexity are more easily observed.[211]

"I propose, therefore, to divide this work into three parts, of which the first shall deal with the conventions necessary for the treatment of the subject, the importance of analogical structures, and the meaning which should be attached to the word species. I will point out on the one hand the evidence of a graduated descending scale, as existing between the highest and the lowest organisms; and, on the other, the effect of surroundings and habits on the organs of living beings, as the cause of their development or arrest of development. Lastly, I will treat of the natural order of animals, and show what should be their fittest classification and arrangement."[212]

It seems unnecessary to give Lamarck's intentions with regard to his second and third parts, as they do not here concern us; they deal with the origin of life and mind.

The first chapter of the work opens with the importance of bearing in mind the difference between the conventional and the natural, that is to say, between words and things. Here, as indeed largely throughout this part of his work, he follows Buffon, by whom he is evidently influenced.

"The conventional deals with systems of arrangement, classification, orders, families, genera, and the nomenclature, whether of different sections or of individual objects.

"An arrangement should be called systematic, or arbitrary, when it does not conform to the genealogical order taken by nature in the development of the things arranged, and when, by consequence it is not founded upon well-considered analogies. There is such a thing as a natural order in every department of nature; it is the order in which its several component items have been successively developed.[213]

"Some lines certainly seem to have been drawn by Nature herself. It was hard to believe that mammals, for example, and birds, were not well-defined classes. Nevertheless the sharpness of definition was an illusion, and due only to our limited knowledge. The ornithorhynchus and the echidna bridge the gulf.[214]

"Simplicity is the main end of any classification. If all the races, or as they are called, species, of any kingdom were perfectly known, and if the true analogies between each species, and between the groups which species form, were also known, so that their approximations to each other and the position of the several groups were in conformity with the natural analogies between them--then classes, orders, sections, and genera would be families, larger or smaller; for each division would be a greater or smaller section of a natural order or sequence.[215] But in this case it would be very difficult to assign the limits of each division; they would be continually subjected to arbitrary alteration, and agreement would only exist where plain and palpable gaps were manifest in our series. Happily, however, for classifiers there are, and will always probably remain, a number of unknown forms."[216]

That the foregoing is still felt to be true by those who accept evolution, may be seen from the following passage, taken from Mr. Darwin's 'Origin of Species':--

"As all the organic beings which have ever lived can be arranged within a few great classes; and as all within each class have, according to our theory, been connected together by fine gradations, the best, and if our collections were nearly perfect, the only possible arrangement would be genealogical:

descent being the hidden bond of connection which naturalists have been seeking under the term of the Natural System. On this view, we can understand how it is that in the eyes of most naturalists, the structure of the embryo is even more important for classifications than that of the adult."[217]

In his second chapter Lamarck deals with the importance of comparative anatomy, and the study of homologous structures. These indicate a sort of blood relationship between the individuals in which they are found, and are our safest guide to any natural system of classification. Their importance is not confined to the study of classes, families, or even species; they must be studied also in the individuals of each species, as it is thus only, that we can recognize either identity or difference of species. The results arrived at, however, are only trustworthy over a limited period, for though the individuals of any species commonly so resemble one another at any given time, as to enable us to generalize from them, at the date of our observing them, yet species are not fixed and immutable through all time: they change, though with such extreme slowness that we do not observe their doing so, and when we come upon a species that *has* changed, we consider it as a new one, and as having always been such as we now see it.[218]

"It is none the less true that when we compare the same kind of organs in different individuals, we can quickly and easily tell whether they are very like each other or not, and hence, whether the animals or plants in which they are found, should be set down as members of the same or of a different species. It is only therefore the general inference drawn from the apparent immutability of species, that has been too inconsiderately drawn.[219]

"The analogies and points of agreement between living organisms, are always incomplete when based upon the consideration of any single organ only. But though still incomplete, they will be much more important according as the organ on which they are founded is an essential one or otherwise.

"With animals, those analogies are most important which exist between organs most necessary for the conservation of their life. With plants, between their organs of generation. Hence, with animals, it will be the interior structure which will determine the most important analogies: with plants it will be the manner in which they fructify.[220]

"With animals we should look to nerves, organs of respiration, and those of the circulation; with plants, to the embryo and its accessories, the sexual organs of their flowers, &c.[221] To do this, will set us on to the Natural Method, which is as it were a sketch traced by man of the order taken by Nature in her productions.[222] Nevertheless the divisions which we shall be obliged to establish, will still be arbitrary and artificial, though presenting to our view sections arranged in the order which Nature has pursued.[223]

"What, then," he asks,[224] "*is* species--and can we show that species has changed--however slowly?" He now covers some of the ground since enlarged upon in Mr. Darwin's second chapter, in which the arbitrary nature of the distinction between species and varieties is so well exposed. "I shall show," says Lamarck (in substance, but I am compelled to condense much), "that the habits by which we now recognize any species, are due to the conditions of life [*circonstances*] under which it has for a long time existed, and that these habits have had such an influence upon the structure of each individual of the species, as to have at length modified this structure, and adapted it to the habits which have been contracted.[225]

"The individuals of any species," he continues, "certainly resemble their parents; it is a universal law of nature that all offspring should differ but little from its immediate progenitors, but this does not justify the ordinary belief that species never vary. Indeed, naturalists themselves are in continual difficulty as regards distinguishing species from varieties; they do not recognize the fact that species are only constant as long as the conditions in which they are placed are constant. Individuals vary and form breeds which blend so insensibly into the neighbouring species, that the distinctions made by naturalists between species and varieties, are for the most part arbitrary, and the confusion upon this head is becoming day by day more serious.[226]

"Not perceiving that species will not vary as long as the conditions in which they are placed remain essentially unchanged, naturalists have supposed that each species was due to a special act of creation on the part of the Supreme Author of all things. Assuredly, nothing can exist but by the will of this Supreme Author, but can we venture to assign rules to him in the execution of his will? May not his infinite power have chosen to create an order of things which should evolve in succession all that we know as well as all that we do not know? Whether we regard species as created or evolved, the boundlessness of his power remains unchanged, and incapable of any diminution whatsoever. Let us then confine ourselves simply to observing the facts around us, and if we find any clue to the path taken by Nature, let us say fearlessly that it has pleased her Almighty Author that she should take this path.[227]

"What applies to species applies also to genera; the further our knowledge extends, the more difficult do we find it to assign its exact limits to any genus. Gaps in our collections are being continually filled up, to the effacement of our dividing lines of demarcation. We are thus compelled to settle the limits of species and variety arbitrarily, and in a manner about which there will be constant disagreement. Naturalists are daily classifying new species which blend into one another so insensibly that there can hardly be found words to express the minute differences between them. The gaps that exist are simply due to our not having yet found the connecting species.

"I do not, however, mean to say that animal life forms a simple and continuously blended series. Life is rather comparable to a ramification. In life we should see, as it were, a ramified continuity, if certain species had not been lost. The species which, according to this illustration, stands at the extremity of each bough, should bear a resemblance, at least upon one side, to the other neighbouring species; and this certainly is what we observe in nature.

"Having arranged living forms in such an order as this, let us take one, and then, passing over several boughs, let us take another at some distance from it; a wide difference will now be seen between the species which the forms selected represent. Our earliest collections supplied us with such distantly allied forms only; now, however, that we have such an infinitely greater number of specimens, we can see that many of them blend one into the other without presenting noteworthy differences at any step."[228]

This has been well extended by Mr. Darwin in a passage which begins:--"The affinities of all beings of the same class have sometimes been represented by a great tree. I believe that this simile largely speaks the truth."[229]

"What, then," continues Lamarck, "can be the cause of all this? Surely the following: namely, that when individuals of any species change their situation, climate, mode of existence, or habits [conditions of life], their structure, form, organization, and in fact their whole being becomes little by little modified, till in the course of time it responds to the changes experienced by the creature."[230]

In his preface Lamarck had already declared that "the thread which gives us a clue to the causes of the various phenomena of animal organization, in the manifold diversity of its developments, is to be found in the fact that Nature conserves in offspring all that their life and environments has developed in parents." Heredity--"the hidden bond of common descent"--tempered with the modifications induced by changed habits--which changed habits are due to new conditions and surroundings--this with Lamarck, as with Buffon and Dr. Darwin, is the explanation of the diversity of forms which we observe in nature. He now goes on to support this--briefly, in accordance with his design--but with sufficient detail to prevent all possibility of mistake about his meaning.

"In the same climate differences in situation, and a greater or less degree of exposure, affect simply, in the first instance, the individuals exposed to them; but in the course of time, these repeated differences of surroundings in individuals which reproduce themselves continually under similar circumstances, induce differences which become part of their very nature; so that after many successive generations, these individuals, which were originally, we will say, of any given species, become transformed into a different one."[231]

"Let us suppose that a grass growing in a low-lying meadow gets carried by some accident to the brow of a neighbouring hill, where the soil is still damp enough for the plant to be able to exist. Let it live here for many generations, till it has become thoroughly accustomed to its position, and let it then gradually find its way to the dry and almost arid soil of a mountain side; if the plant is able to stand the change and to perpetuate itself for many generations, it will have become so changed that botanists will class it as a new species."[232]

"The same sort of process goes on in the animal kingdom, but animals are modified more slowly than plants."[233]

The sterility of hybrids, to which Mr. Darwin devotes a great part of the ninth chapter of his 'Origin of Species,'[234] is then touched on--briefly, but sufficiently--as follows:--

"The idea that species were fixed and immutable involved the belief that distinct species could not be fertile *inter se*. But unfortunately observation has proved, and daily proves, that this supposition is unfounded. Hybrids are very common among plants, and quite sufficiently so among animals to show that the boundaries of these so-called immutable species are not so well defined as has been supposed. Often, indeed, there is no offspring between the individuals of what are called distinct species, especially when they are widely different, and again, the offspring when produced is generally sterile; but when there is less difference between the parents, both the difficulty of breeding the hybrid, and its sterility when produced, are found to disappear. In this very power of crossing we see a source from which breeds, and ultimately species, may arise."[235]

Mr. Darwin arrives at the same conclusion. He writes:--

"We must, therefore, either give up the belief of the universal sterility of species when crossed, or we must look at this sterility in animals, not as an indelible characteristic, but as one capable of being removed by domestication.

"Finally, on considering all the ascertained facts on the intercrossing of plants and animals, it may be concluded that some degree of sterility, both in first crosses and in hybrids, is an exceedingly general result, but that it cannot, under our present state of knowledge, be considered as absolutely universal."[236]

Returning to Lamarck, we find him saying:--

"The limits, therefore, of so-called species are not so constant and unvarying as is commonly supposed. Consider also the following. All living forms upon the face of the globe have been brought forth in the course of infinite time by the process of generation only. Nature has directly created none but the lowest organisms; these she is still producing every day, they being, as it were, the first sketches of life, and produced by what is called spontaneous generation. Organs have been gradually developed in these low forms, and these organs have in the course of time increased in diversity and complexity. The power of growth in each living body has given rise to various modes of reproduction, and thus progress, already acquired, has been preserved and handed down to offspring.[237] With sufficient time, favourable conditions of life [*circonstances*], successive changes in the surface of the globe, and the power of new surroundings and habits to modify the organs of living bodies, all animal and vegetable forms have been imperceptibly rendered such as we now see them. It follows that species will be constant only in relation to their environments, and cannot be as old as Nature herself.

"But what are we to say of instinct? Can we suppose that all the tricks, cunning, artifices, precautions, patience, and skill of animals are due to evolution only? Must we not see here the design of an all-powerful Creator? No one certainly will assign limits to the Creator's power, but it is a bold thing to say that he did not choose to work in this way or that way, when his own handiwork declares to us that this is the way he chose. I find proof in Nature--meaning by nature the *ensemble* of all that is,[238] but regarding her as herself the effect of an unknown first cause[239]--that she is the author of organization, life, and even sensation; that she has multiplied and diversified the organs and mental powers of the creatures which she sustains and reproduces; that she has developed in animals, through the sole instrumentality of sense of need as establishing and directing their habits, all actions and all habits, from the simplest up to those which constitute instinct, industry, and finally reason.[240]

"Against this it is alleged that we have no reason to believe species to have changed within any known era. The skeletons of some Egyptian birds, preserved two or three thousand years ago, differ in no particular from the same kind of creatures at the present day. But this is what we should expect, inasmuch as the position and climate of Egypt itself do not appear to have changed. If the conditions of life have not varied, why should the species subjected to those conditions have done so? Moreover, birds can move about freely, and if one place does not suit them they can find another that does. All that these Egyptian mummies really prove is, that there were animals in Egypt two or three thousand years ago which are like the animals of to-day; but how short a space is two or three thousand years, as compared with the time which Nature has had at her disposal! A time infinitely great *quâ* man, is still infinitely short *quâ* Nature.[241]

"If, however, we turn to animals under confinement, we find immediate proof that the most startling changes are capable of being produced after some generations of changed habits. In the sixth chapter we shall have occasion to observe the power of changed conditions [*circonstances*] to develop new desires in animals, and to induce new courses of action; we shall see the power which these new actions will have, after a certain amount of repetition, to engender new habits and tendencies; and we shall also note the effects of use and disuse in either fortifying and developing an organ, or in diminishing it and causing it to disappear. With plants under domestication, we shall find corresponding phenomena. Species will thus appear to be unchangeable for comparatively short periods only."[242]

It is interesting to see that Mr. Darwin lays no less stress on the study of animals and plants under domestication than Buffon, Dr. Darwin, and Lamarck. Indeed, all four writers appear to have been in great measure led to their conclusions by this very study. "At the commencement of my investigations," writes Mr. Darwin, "it seemed to me probable that a careful study of domesticated animals and of cultivated plants would offer the best chance of making out this obscure problem. Nor have I been disappointed; in this and in all other perplexing cases, I have invariably found that our knowledge, imperfect though it be, of variation under domestication, afforded the best and safest clue. I may venture to express my conviction of the high value of such studies, though they have been very commonly neglected by naturalists."[243]

In justice to the three writers whom I have named, it should be borne in mind that they also ventured to express their conviction of the high value of these studies. Buffon, indeed, as we have seen, gives animals under domestication the foremost place in his work. He does not treat of wild animals till he has said all he has to say upon our most important domesticated breeds,--on whose descent from one or two wild stocks he is never weary of insisting. It was doubtless because of the opportunities they afforded him for demonstrating the plasticity of living organism that the most important position in his work was assigned to them.

Lamarck professes himself unable to make up his mind about extinct species; how far, that is to say, whole breeds must be considered as having died out, or how far the difference between so many now living and fossil forms is due to the fact that our living species are modified descendants of the fossil ones. Such large parts of the globe were still practically unknown in Lamarck's time, and the recent discovery of the ornithorhynchus has raised such hopes as to what might yet be found in Australia, that he was inclined to think that only such creatures as man found hurtful to him, as, for example, the megatherium and the mastodon, had become truly extinct, nor was he, it would seem, without a hope that these would yet one day be discovered. The climatic and geological changes that have occurred in past ages, would, he believed, account for all the difference which we observe between living and fossil forms, inasmuch as they would have changed the conditions under which animals lived, and therefore their habits and organs would have become correspondingly modified. He therefore rather wondered to find so much, than so little, resemblance between existing and fossil forms.

Buffon took a juster view of this matter; it will be remembered that he concluded his remarks upon the mammoth by saying that many species had doubtless disappeared without leaving any living descendants, while others had left descendants which had become modified.

Lamarck anticipated Lyell in supposing geological changes to have been due almost entirely to the continued operation of the causes which we observe daily at work in nature: thus he writes:--

"Every observer knows that the surface of the earth has changed; every valley has been exalted, the crooked has been made straight, and the rough places plain; not even is climate itself stable. Hence changed conditions; and these involve changed needs and habits of life; if such changes can give rise to modifications or developments, it is clear that every living body must vary, especially in its outward character, though the variation can only be perceptible after several generations.

"It is not surprising then that so few living species should be represented in the geologic record. It is surprising rather that we should find any living species represented at all.[244]

"Catastrophes have indeed been supposed, and they are an easy way of getting out of the difficulty; but unfortunately, they are not supported by evidence. Local catastrophes have undoubtedly occurred, as earthquakes and volcanic eruptions, of which the effects can be sufficiently seen; but why suppose any universal catastrophe, when the ordinary progress of nature suffices to account for the phenomena? Nature is never *brusque*. She proceeds slowly step by step, and this with occasional local catastrophes will remove all our difficulties."[245]

In his fourth chapter Lamarck points out that animals move themselves, or parts of themselves, not through impulsion or movement communicated to them as from one billiard ball to another, but by reason of a cause which excites their irritability, which cause is within some animals and forms part of them, while it is wholly outside of others.[246]

I should again warn the reader to be on his guard against the opinion that any animals can be said to live if they have no "inward motion" of their own which prompts them to act. We cannot call anything alive which moves only as wind and water may make it move, but without any impulse from within to execute the smallest action and without any capacity of feeling. Such a creature does not look sufficiently like the other things which we call alive; it should be first shown to us, so that we may make up our minds whether the facts concerning it have been truly stated, and if so, what it most resembles; we may then classify it accordingly.

"Some animals change their place by creeping, some by walking, some by running or leaping; others again fly, while others live in the water and swim.

"The origin of these different kinds of locomotion is to be found in the two great wants of animal life: 1, the means of procuring food; 2, the search after mates with a view to reproduction.

"Since then the power of locomotion was a matter affecting their individual self-preservation, as well as that of their race, the existence of the want led to the means of its being gratified."[247]

Lamarck is practically at one with Dr. Erasmus Darwin, that modification will commonly travel along three main lines which spring from the need of reproduction, of procuring food, and (Dr. Darwin has added) the power of self-protection; but Dr. Darwin's treatment of this part of his subject is more lucid and satisfactory than Lamarck's, inasmuch as he immediately brings forward instances of various modifications which have in each case been due to one of the three main desires above specified, namely, reproduction, subsistence, and self-defence.

Lamarck concludes the chapter with some passages which show that he was alive--as what Frenchman could fail to be after Buffon had written?--to the consequences which must follow from the geometrical ratio of increase, and to the struggle for existence, with consequent survival of the fittest, which must always be one of the conditions of any wild animal's existence. The paragraphs, indeed, on this subject are taken with very little alteration from Buffon's work. As Lamarck's theory is based upon the fact that it is on the nature of these conditions that the habits and consequently the structure of any animal will depend, he must have seen that the shape of many of its organs must vary greatly in correlation to the conditions to which it was subjected in the matter of self-protection. I do not see, then, that there is any substantial difference between the positions taken by Dr. Erasmus Darwin and by Lamarck in this respect.

"Let us conclude," he writes, "by showing the means employed by nature to prevent the number of her creatures from injuring the conservation of what has been produced already, and of the general order which should subsist.[248]

"In consequence of the extremely rapid rate of increase of the smaller, and especially of the most imperfect, animals, their numbers would become so great as to prove injurious to the conservation of breeds, and to the progress already made towards more perfect organization, unless nature had taken precautions to keep them down within certain fixed limits which she cannot exceed."[249]

This seems to contain, and in a nutshell, as much of the essence of what Mr. Herbert Spencer and Mr. Charles Darwin have termed the survival of the fittest in the struggle for existence, as was necessary for Lamarck's purpose.

To Lamarck, as to Dr. Darwin and Buffon, it was perfectly clear that the facts, that animals have to find their food under varying circumstances, and that they must defend themselves in all manner of varying ways against other creatures which would eat them if they could, were simply some of the conditions of their existence. In saying that the surrounding circumstances--which amount to the conditions of existence--determined the direction in which any plant or animal should be slowly modified, Lamarck includes as a matter of course the fact that the "stronger and better armed should eat the weaker," and thus survive and bear offspring which would inherit the strength and better armour of its parents. Nothing therefore can be more at variance with the truth than to represent Lamarck and the other early evolutionists as ignoring the struggle for existence and the survival of the fittest; these are inevitably implied whenever they use the word "*circonstances*" or environment, as I will more fully show later on, and are also expressly called attention to by the greater number of them.[250]

"Animals, except those which are herbivorous, prey upon one another; and the herbivorous are exposed to the attacks of the flesh-eating races.

"*The strongest and best armed for attack eat the weaker*, and the greater kinds eat the smaller. Individuals of the same race rarely eat one another; they war only with other races than their own."[251]

Dr. Darwin here again has the advantage over Lamarck; for he has pointed out how the males contend with one another for the possession of the females, which I do not find Lamarck to have done, though he would at once have admitted the fact. Lamarck continues:--

"The smaller kinds of animals breed so numerously and so rapidly that they would people the globe to the exclusion of other forms of life, if nature had not limited their inconceivable multitude. As, however, they are the prey of a number of other creatures, live but a short time, and perish easily with cold, they are kept always within the proportions necessary for the maintenance both of their own and of other races.[252]

"As regards the larger and stronger animals, they would become dominant, and be injurious to the conservation of many other races, if they could multiply in too great numbers. But as it is, they devour one another, and breed but slowly, and few at a birth, so that equilibrium is duly preserved among them. Man alone is the unquestionably dominant animal, but men war among themselves, so that it may be safely said the world will never be peopled to its utmost capacity."[253]

In his fifth chapter Lamarck returns to the then existing arrangement and classification of animals.

"Naturalists having remarked that many species, and some genera and even families present characters which as it were isolate them, it has been imagined that these approached or drew further from each other according as their points of agreement or difference seemed greater or less when set down as it were on a chart or map. They regard the small well-marked series which have been styled natural families, as groups which should be placed between the isolated species and their nearest neighbours so as to form a kind of reticulation. This idea, which some of our modern naturalists have held to be admirable, is evidently mistaken, and will be discarded on a profounder and more extended knowledge of organization, and more especially when the distinction has been duly drawn between what is due to the action of special conditions and to general advance of organization."[254]

I take it that Lamarck is here attempting to express what Mr. Charles Darwin has rendered much more clearly in the following excellent passage:--

"It should always be borne in mind what sort of intermediate forms must, on the theory [what theory?], have formerly existed. I have found it difficult when looking at any two species to avoid picturing to myself forms *directly* intermediate between them. But this is a wholly false view; we should always look for forms intermediate between each species and a common but unknown progenitor; and the progenitor will generally have differed in some respects from all its modified descendants. To give a simple illustration: the fantail and pouter pigeons are both descended from the rock pigeon. If we possessed all the intermediate varieties which have ever existed, we should have an extremely close series, between both and the rock pigeon; but we should have no varieties directly intermediate between the fantail and the pouter; none, for instance, combining a tail somewhat expanded with a crop somewhat enlarged, the characteristic features of these two breeds. These two breeds, moreover, have become so much modified that, if we had no historical or indirect evidence regarding their origin, it would not have been possible to have determined, from a mere comparison of their structure with that of the rock pigeon C. livia, whether they had descended from this species, or from some other allied form, as C. oenas.

"So with natural species, if we look to forms very distinct--for instance, to the horse and the tapir--we have no reason to suppose that links directly intermediate between them ever existed, but between each and an unknown common parent. The common parent will have had in its whole organization much general resemblance to the tapir and the horse; but in some points of structure it may have differed considerably from both, even perhaps more than they differ from each other. Hence in all such cases we should be unable to recognize the parent form of any two or more species, even if we closely compared the structure of the parent with that of its modified descendants, unless at the same time we had a nearly perfect chain of the intermediate links.

"By the theory of natural selection [surely this is a slip for "by the theory of descent with modification"] all living species have been connected with the parent species of each genus, by differences not greater than we see between the natural and domestic varieties of the same species at the present day; and their parent species, now generally extinct, have in their turn been similarly connected with more ancient forms, and so on backwards, always converging to the common ancestor of each great class; so that the number of intermediate and transitional links between all living and extinct species must have been inconceivably great. But assuredly if this theory [the theory of descent with modification or that of "natural selection"?] be true, such have lived upon the earth."[255]

To return, however, to Lamarck.

"Though Nature," he continues, "in the course of long time has evolved all animals and plants in a true scale of progression, the steps of this scale can be perceived only in the principal groups of living forms; it cannot be perceived in species nor even in genera. The reason of this lics in the extreme diversity of the surroundings in which each different race of animals and plants has existed. These surroundings have often been out of harmony with the growing organization of the plants and animals themselves; this has led to anomalies, and, as it were, digressions, which the mere development of organization by itself could not have occasioned."[256] Or, in other words, to that divergency of type which is so well insisted on by

Mr. Charles Darwin.

"It is only therefore the principal groups of animal and vegetable life which can be arranged in a vertical line of descent; species and even genera cannot always be so--for these contain beings whose organization has been dependent on the possession of such and such a special system of essential organs.

"Each great and separate group has its own system of essential organs, and it is these systems which can be seen to descend, within the limits of the group, from their most complex to their simplest form. But each organ, considered individually, does not descend by equally regular gradation; the gradations are less and less regular according as the organ is of less importance, and is more susceptible of modification by the conditions which surround it. Organs of small importance, and not essential to existence, are not always either perfected or degraded at an equal rate, so that in observing all the species of any class we find an organ in one species in the highest degree of perfection, while another organ, which in this same species is impoverished or very imperfect, is highly developed in another species of the same group."[257]

The facts maintained in the preceding paragraph are in great measure supported by Mr. Charles Darwin, who, however, assigns their cause to natural selection.

Mr. Darwin writes, "Ordinary specific characters are more variable than generic;" and again, a little lower down, "The points in which all the species of a genus resemble each other, and in which they differ from allied genera, are called generic characters; and these characters may be attributed to inheritance from a common progenitor, for it can rarely happen that natural selection will have modified several distinct species fitted to more or less widely different habits, in exactly the same manner; and as these so called generic characters have been inherited from before the period when the several species first branched off from their common progenitor, and subsequently have not varied or come to differ in any degree, or only in a slight degree, it is not probable that they should vary at the present day. On the other hand, the points in which species differ from other species of the same genus are called specific characters; and as these specific characters have varied and come to differ since the period when the species branched off from a common progenitor, it is probable that they should still often be in some degree variable, or at least more variable than those parts of the organization which have for a very long time remained constant."[258]

The fact, then, that it is specific characters which vary most is agreed upon by both Lamarck and Mr. Darwin. Lamarck, however, maintains that it is these specific characters which are most capable of being affected by the habits of the creature, and that it is for this reason they will be most variable, while Mr. Darwin simply says they *are* most variable, and that, this being so, the favourable variations will be preserved and accumulated--an assertion which Lamarck would certainly not demur to.

"Irregular degrees of perfection," says Lamarck, "and degradation in the less essential organs, are due to the fact that these are more liable than the more essential ones to the influence of external circumstances: these induce corresponding differences in the more outward parts of the animal, and give rise to such considerable and singular difference in species, that instead of being able to arrange them in a direct line of descent, as we can arrange the main groups, these species often form lateral ramifications round about the main groups to which they belong, and in their extreme development are truly isolated."[259]

In his summary of the second chapter of his 'Origin of Species,' Mr. Darwin well confirms this when he says, "In large genera the species are apt to be closely, but unequally, allied together, forming little clusters round other species."

"A longer time," says Lamarck, "and a greater influence of surrounding conditions, is necessary in order to modify interior organs. Nevertheless we see that Nature does pass from one system to another without any sudden leap, when circumstances require it, provided the systems are not too far apart. Her method is to proceed from the more simple to the more complex.[260]

"She does this not only in the race, but in the individual." Here Lamarck, like Dr. Erasmus Darwin, shows his perception of the importance of embryology in throwing light on the affinities of animals--as since more fully insisted on by the author of the 'Vestiges of Creation,' and by Mr.

Darwin,[261] as well as by other writers. "Breathing through gills is nearer to breathing through lungs than breathing through trachea is. Not only do we see Nature pass from gills to lungs in families which are not too far apart, as may be seen by considering the case of fishes and reptiles; but she does so during the existence of a single individual, which may successively make use both of the one and of the other system. The frog while yet a tadpole breathes through gills; on becoming a frog it breathes through lungs; but we cannot find that Nature in any case passes from trachea to lungs."[262]

Lamarck now rapidly reviews previous classifications, and propounds his own, which stands thus:--I. Vertebrata, consisting of Mammals, Birds, Fishes, and Reptiles. II. Invertebrata, consisting of Molluscs, Centipedes, Annelids, Crustacea, Arachnids, Insects, Worms, Radiata, Polyps, Infusoria.

"The degradation of organism," he concludes, "in this descending scale is not perfectly even, and cannot be made so by any classification, nevertheless there is such evidence of sustained degradation in the principal groups as must point in the direction of some underlying general principle."[263]

Lamarck's sixth chapter is headed "Degradation and Simplification of the Animal Chain as we proceed downwards from the most complex to the most simple Organisms."

"This is a positive fact, and results from the operation of a constant law of nature; but a disturbing cause, which can be easily recognized, varies the regular operation of the law from one end to the other of the chain of life.[264]

"We can see, nevertheless, that special organs become more and more simple the lower we descend; that they become changed, impoverished, and attenuated little by little; that they lose their local centres, and finally become definitely annihilated before we reach the lowest extremity of the chain.[265]

"As has been said already, the degradation of organism is not always regular; such and such an organ often fails or changes suddenly, and sometimes in its changes assumes forms which are not allied with any others by steps that we can recognize. An organ may disappear and reappear several times before being entirely lost: but this is what we might expect, for the cause which has led to the evolution of living organisms has evolved many varieties, due to external influences. Nevertheless, looking at organization broadly, we observe a descending scale."[266]

"If the tendency to progressive development was the only cause which had influenced the forms and organs of animals, development would have been regular throughout the animal chain; but it has not been so: Nature is compelled to submit her productions to an environment which acts upon them, and variation in environment will induce variation in organism: this is the true cause of the sometimes strange deviations from the direct line of progression which we shall have to observe.[267]

"If Nature had only called aquatic beings into existence, and if these beings had lived always in the same climate, in the same kind of water, and at the same depth, the organization of these animals would doubtless have presented an even and regular scale of development. But there has been fresh water, salt water, running and stagnant water, warm and cold climates, an infinite variety of depth: animals exposed to these and other differences in their surroundings have varied in accordance with them.[268] In like manner those animals which have been gradually fitted for living in air instead of water have been subjected to an endless diversity in their surroundings. The following law, then, may be now propounded, namely:--

"*That anomalies in the development of organism are due to the influences of the environment and to the habits of the creature.* [269]

"Some have said that the anomalies above mentioned are so great as to disprove the existence of any scale which should indicate descent; but the nearer we approach species, the smaller we see differences become, till with species itself we find them at times almost imperceptible."[270]

Lamarck here devotes about seventy pages to a survey of the animal kingdom in its entirety, beginning with the mammals and ending with the infusoria. He points out the manner in which organ after organ disappears as we descend the scale, till we are left with a form which, though presenting all the characteristics of life, has yet no special organ whatever. I am obliged to pass this classification over, but do so very unwillingly, for it is illustrative of Lamarck, both at his best and at his worst.

The seventh chapter is headed--

"On the influence of their surroundings on the actions and habits of animals, and on the effect of these habits and actions in modifying their organization."

"The effect of different conditions of our organization upon our character, tendencies, actions, and even our ideas, has been often remarked, but no attention has yet been paid to that of our actions and habits upon our organization itself. These actions and habits depend entirely upon our relations to the surroundings in which we habitually exist; we shall have occasion, therefore, to see how great is the effect of environment upon organization.

"But for our having domesticated plants and animals we should never have arrived at the perception of this truth; for though the influence of the environment is at all times and everywhere active upon all living bodies, its effects are so gradual that they can only be perceived over long periods of time.[271]

"Taking the chain of life in the inverse order of nature--that is to say, from man downwards--we certainly perceive a sustained but irregular degradation of organism, with an increasing simplicity both in organism and faculties.

"This fact should throw light upon the order taken by nature, but it does not show us why the gradation is so irregular, nor why throughout its extent we find so many anomalies or digressions which have apparently no order at all in their manifold varieties.[272] The explanation of this must be sought for in the infinite diversity of circumstances under which organisms have been developed. On the one hand, there is a tendency to a regular progressive development; on the other, there is a host of widely different surroundings which tend continually to destroy the regularity of development.

"It is necessary to explain what is meant by such expressions as 'the effect of its environment upon the form and organization of an animal.' It must not be supposed that its surroundings directly effect any modification whatever in the form and organization of an animal.[273] Great changes in surroundings involve great changes in the wants of animals, and these changes in their wants involve corresponding changes in their actions. If these new wants become permanent, or of very long duration, the animals contract new habits, which last as long as the wants which gave rise to them.[274] A great change in surroundings, if it persist for a long time, must plainly, therefore, involve the contraction of new habits. These new habits in their turn involve a preference for the employment of such and such an organ over such and such another organ, and in certain cases the total disuse of an organ which is no longer wanted. This is perfectly self-evident.[275]

"On the one hand, new wants have rendered a part necessary, which part has accordingly been created by a succession of efforts: use has kept it in existence, gradually strengthening and developing it till in the end it attains a considerable degree of perfection. On the other, new circumstances having in some cases rendered such or such a part useless, disuse has led to its gradually ceasing to receive the development which the other parts attain to; on this it becomes reduced, and in time disappears.[276]

"Plants have neither actions nor habits properly so called, nevertheless they change in a changed environment as much as animals do. This is due to changes in nutrition, absorption and transpiration, to degrees of heat, light, and moisture, and to the preponderance over others which certain of the vital functions attain to."

Lamarck is led into the statement that plants have neither actions nor habits, by his theories about the nervous system and the brain. Plain matter-of-fact people will prefer the view taken by Buffon, Dr. Darwin, and, more recently, by Mr. Francis Darwin, that there is no radical difference between plants and animals.

"The differences between well-nourished and ill-nourished plants become little by little very noticeable. If individuals, whether animal or vegetable, are continually ill-fed and exposed to hardships for several generations, their organization becomes eventually modified, and the modification is transmitted until a race is formed which is quite distinct from those descendants of the common parent stock which have been placed in favourable circumstances.[277] In a dry spring the meagre and stunted herbage seeds early. When, on the other hand, the spring is warm but with occasional days of rain, there is an excellent hay-crop. If, however, any cause perpetuates unfavourable circumstances, plants will vary correspondingly, first in appearance and general conditions, and then in several particulars of their actual character, certain organs having received more development than others, these differences will in the course of time become hereditary.[278]

"Nature changes a plant or animal's surroundings gradually--man sometimes does so suddenly. All botanists know that plants vary so greatly under domestication that in time they become hardly recognizable. They undergo so much change that botanists do not at all like describing domesticated varieties. Wheat itself is an example. Where can wheat be found as a wild plant, unless it have escaped from some neighbouring cultivation? Where are our cauliflowers, our lettuces, to be found wild, with the same characters as they possess in our kitchen gardens?

"The same applies to our domesticated breeds of animals. What a variety of breeds has not man produced among fowls and pigeons, of which we can find no undomesticated examples!"[279]

The foregoing remarks on the effects of

domestication seem to have been inspired by those given p. 123 and pp. 168, 169 of this volume.[280]

"Some, doubtless, have changed less than others, owing to their having undergone a less protracted domestication, and a less degree of change in climate; nevertheless, though our ducks and geese, for example, are of the same type as their wild progenitors, they have lost the power of long and sustained flight, and have become in other respects considerably modified.[281]

"A bird, after having been kept five or six years in a cage, cannot on being liberated fly like its brethren which have been always free. Such a change in a single lifetime has not effected any transmissible modification of type; but captivity, continued during many successive generations, would undoubtedly do so. If to the effects of captivity there be added also those of changed climate, changed food, and changed actions for the purpose of laying hold of food, these, united together and become constant, would in the course of time develop an entirely new breed."

This, again, is almost identical with the passage from Buffon,[282] p. 148 of this volume. See also pp. 169, 170.

"Where can our many domestic breeds of dogs be found in a wild state? Where are our bulldogs, greyhounds, spaniels, and lapdogs, breeds presenting differences which, in wild animals, would be certainly called specific? These are all descended from an animal nearly allied to the wolf, if not from the wolf itself. Such an animal was domesticated by early man, taken at successive intervals into widely different climates, trained to different habits, carried by man in his migrations as a precious capital into the most distant countries, and crossed from time to time with other breeds which had been developed in similar ways. Hence our present multiform breeds."[283]

Here, also, it is impossible to forget Buffon's passages on the dog, given pp. 121, 122. See also p. 223.

"Observe the gradations which are found between the *ranunculus aquatilis* and the *ranunculus hederaceus*: the latter--a land plant--resembles those parts of the former which grow above the surface of the water, but not those that grow beneath it.[284]

"The modifications of animals arise more slowly than those of plants; they are therefore less easily watched, and less easily assignable to their true causes, but they arise none the less surely. As regards these causes, the most potent is diversity of the surroundings in which they exist, but there are also many others.[285]

"The climate of the same place changes, and the place itself changes with changed climate and exposure, but so slowly that we imagine all lands to be stable in their conditions. This, however, is not true; climatic and other changes induce corresponding changes in environment and habit, and these modify the structure of the living forms which are subjected to them. Indeed, we see intermediate forms and species corresponding to intermediate conditions.

"To the above causes must be ascribed the infinite variety of existing forms, independently of any tendency towards progressive development."[286]

The reader has now before him a fair sample of "the well-known doctrine of inherited habit as advanced by Lamarck."[287] In what way, let me ask in passing, does "the case of neuter insects" prove "demonstrative" against it, unless it is held equally demonstrative against Mr. Darwin's own position? Lamarck continues:--

"The character of any habitable quarter of the globe is *quâ* man constant: the constancy of type in species is therefore also *quâ* man persistent. But this is an illusion. We establish, therefore, the three following propositions:--

"1. That every considerable and sustained change in the surroundings of any animal involves a real change in its needs.

"2. That such change of needs involves the necessity of changed action in order to satisfy these needs, and, in consequence, of new habits.[288]

"3. It follows that such and such parts, formerly less used, are now more frequently employed, and in consequence become more highly developed; new parts also become insensibly evolved in the creature by its own efforts from within.

"From the foregoing these two general laws may be deduced:--

"*Firstly. That in every animal which has not passed its limit of development, the more frequent and sustained employment of any organ develops and aggrandizes it, giving it a power proportionate to the duration of its employment, while the same organ in default of constant use becomes insensibly weakened and deteriorated, decreasing imperceptibly in power until it finally disappears.*[289]

"*Secondly. That these gains or losses of organic development, due to use or disuse, are transmitted to offspring, provided they have been common to both sexes, or to the animals from which the offspring have descended.*"[290]

Lamarck now sets himself to establish the fact that animals have developed modifications which have been transmitted to their offspring.

"Naturalists," he says, "have believed that the possession of certain organs has led to their employment. This is not so: it is need and use which have developed the organs, and even called them into existence." [I have already sufficiently insisted that it is impossible to dispense with either of these two views. Demand and Supply have gone hand in hand, each reacting upon the other.] "Otherwise a special act of creation would be necessary for every different combination of conditions; and it would be also necessary that the conditions should remain always constant.

"If this were really so we should have no racehorses like those of England, nor drayhorses so heavy in build and so unlike the racehorse; for there are no such breeds in a wild state. For the same reason, we should have no turnspit dogs with crooked legs, no greyhounds nor water-spaniels; we should have no tailless breed of fowls nor fantail pigeons, &c. Nor should we be able to cultivate wild plants in our gardens, for any length of time we please, without fear of their changing.

"'Habit,' says the proverb, 'is a second nature'; what possible meaning can this proverb have, if descent with modification is unfounded?[291]

"As regards the circumstances which give rise to variation, the principal are climatic changes, different temperatures of any of a creature's environments, differences of abode, of habit, of the most frequent actions; and lastly, of the means of obtaining food, self-defence, reproduction, &c., &c."[292]

Here we have absolute agreement with Dr. Erasmus Darwin,[293] except that there seems a tendency in this passage to assign more effect to the direct action of conditions than is common with Lamarck. He seems to be mixing Buffon and Dr. Darwin.

"In consequence of change in any of these respects, the faculties of an animal become extended and enlarged by use: they become diversified through the long continuance of the new habits, until little by little their whole structure and nature, as well as the organs originally affected, participate in the effects of all these influences, and are modified to an extent which is capable of transmission to offspring."[294]

This sentence alone would be sufficient to show that Lamarck was as much alive as Buffon and Dr. Darwin were before him, to the fact that one of the most important conditions of an animal's life, is the relation in which it stands to the other inhabitants of the same neighbourhood--from which the survival of the fittest follows as a self-evident proposition. Nothing, therefore, can be more unfounded than the attempt, so frequently made by writers who have not read Lamarck, or who think others may be trusted not to do so, to represent him as maintaining something perfectly different from what is maintained by modern writers on evolution. The difference, in so far as there is any difference, is one of detail only. Lamarck would not have hesitated to admit, that, if animals are modified in a direction which is favourable to them, they will have a better chance of surviving and transmitting their favourable modifications. In like manner, our modern evolutionists should allow that animals are modified not because they subsequently survive, but because they have done this or that which has led to their modification, and hence to their surviving.

Having established that animals and plants are capable of being materially changed in the course of a few generations, Lamarck proceeds to show that their modification is due to changed distribution of the use and disuse of their organs at any given time.

"*The disuse of an organ*," he writes, "*if it becomes*

constant in consequence of new habits, gradually reduces the organ, and leads finally to its disappearance."[295]

"Thus whales have lost their teeth, though teeth are still found in the embryo. So, again, M. Geoffroy has discovered in birds the groove where teeth were formerly placed. The ant-eater, which belongs to a genus that has long relinquished the habit of masticating its food, is as toothless as the whale."[296]

Then are adduced further examples of rudimentary organs, which will be given in another place, and need not be repeated here. Speaking of the fact, however, that serpents have no legs, though they are higher in the scale of life than the batrachians, Lamarck attributes this "to the continued habit of trying to squeeze through very narrow places, where four feet would be in the way, and would be very little good to them, inasmuch as more than four would be wanted in order to turn bodies that were already so much elongated."[297]

If it be asked why, on Lamarck's theory, if serpents wanted more legs they could not have made them, the answer is that the attempt to do this would be to unsettle a question which had been already so long settled, that it would be impossible to reopen it. The animal must adapt itself to four legs, or must get rid of all or some of them if it does not like them; but it has stood so long committed to the theory that if there are to be legs at all, there are to be not more than four, that it is impossible for it now to see this matter in any other light.

The experiments of M. Brown Séquard on guinea pigs, quoted by Mr. Darwin,[298] suggest that the form of the serpent may be due to its having lost its legs by successive accidents in squeezing through narrow places, and that the wounds having been followed by disease, the creature may have bitten the limbs off, in which case the loss might have been very readily transmitted to offspring; the animal would accordingly take to a sinuous mode of progression that would doubtless in time elongate the body still further. M. Brown Séquard "carefully recorded" thirteen cases, and saw even a greater number, in which the loss of toes by guinea pigs which had gnawed their own toes off, was immediately transmitted to offspring. Accidents followed by disease seem to have been somewhat overlooked as a possible means of modification. The missing forefinger to the hand of the potto[299] would appear at first sight to have been lost by some such mishap. Returning to Lamarck, we find him saying:--

"Even in the lifetime of a single individual we can see organic changes in consequence of changed habits. Thus M. Tenon has constantly found the intestinal canal of drunkards to be greatly shorter than that of people who do not drink. This is due to the fact that habitual drunkards eat but little solid food, so that the stomach and intestines are more rarely distended. The same applies to people who lead studious and sedentary lives. The stomachs of such persons and of drunkards have little power, and a small quantity will fill them, while those of men who take plenty of exercise remain in full vigour and are even increased."[300]

It becomes now necessary to establish the converse proposition, namely that:--

"*The frequent use of an organ increases its power; it even develops the organ itself, and makes it acquire dimensions and powers which it is not found to have in animals which make no use of such an organ.*

"In support of this we see that the bird whose needs lead it to the water, in which to find its prey, extends the toes of its feet when it wants to strike the water, and move itself upon the surface. The skin at the base of the toes of such a bird contracts the habit of extending itself from continual practice. To this cause, in the course of time, must be attributed the wide membrane which unites the toes of ducks, geese, &c. The same efforts to swim, that is to say, to push the water for the purpose of moving itself forward, has extended the membrane between the toes of frogs, turtles, the otter, and the beaver."[301]

[This is taken, I believe, from Dr. Darwin or Buffon, but I have lost the passage, if, indeed, I ever found it. It had been met by Paley some years earlier (1802) in the following:--

"There is nothing in the action of swimming as carried on by a bird upon the surface of the water that should generate a membrane between the toes. As to that membrane it is an action of constant resistance.... The web feet of amphibious quadrupeds, seals, otters, &c., fall under the same observation."[302]]

"On the other hand those birds whose habits lead them to perch on trees, and which have sprung from parents that have long contracted this habit, have their toes shaped in a perfectly different manner. Their claws become lengthened, sharpened, and curved, so as to enable the creature to lay hold of the boughs on which it so often rests. The shore bird again, which does not like to swim, is nevertheless continually obliged to enter the water when searching after its prey. Not liking to plunge its body in the water, it makes every endeavour to extend and lengthen its lower limbs. In the course of long time these birds have come to be elevated, as it were, on stilts, and have got long legs bare of feathers as far as their thighs, and often still higher. The same bird is continually trying to extend its neck in order to fish without wetting its body, and in the course of time its neck has become modified accordingly.[303]

"Swans, indeed, and geese have short legs and very long necks, but this is because they plunge their heads as low in the water as they can in their search for aquatic larvæ and other animalcules, but make no effort to lengthen their legs."[304]

This too is taken from some passage which I have either never seen or have lost sight of. Paley never gives a reference to an opponent, though he frequently does so when quoting an author on his own side, but I can hardly doubt that he had in his mind the passage from which Lamarck in 1809 derived the foregoing, when in 1802 he wrote § 5 of chapter xv. and the latter half of chapter xxiii. of his 'Natural Theology.'

"The tongues of the ant-eater and the woodpecker," continues Lamarck, "have become elongated from similar causes. Humming birds catch hold of things with their tongues; serpents and lizards use their tongues to touch and reconnoitre objects in front of them, hence their tongues have come to be forked.

"Need--always occasioned by the circumstances in which an animal is placed, and followed by sustained efforts at gratification--can not only modify an organ, that is to say, augment or reduce it, but can change its position when the case requires its removal.[305]

"Ocean fishes have occasion to see what is on either side of them, and have their eyes accordingly placed on either side their head. Some fishes, however, have their abode near coasts on submarine banks and inclinations, and are thus forced to flatten themselves as much as possible in order to get as near as they can to the shore. In this situation they receive more light from above than from below, and find it necessary to pay attention to whatever happens to be above them; this need has involved the displacement of their eyes, which now take the remarkable position which we observe in the case of soles, turbots, plaice, &c. The transfer of position is not even yet complete in the case of these fishes, and the eyes are not, therefore, symmetrically placed; but they are so with the skate, whose head and whole body are equally disposed on either side a longitudinal section. Hence the eyes of this fish are placed symmetrically upon the uppermost side.[306]

"The eyes of serpents are placed on the sides and upper portions of the head, so that they can easily see what is on one side of them or above them; but they can only see very little in front of them, and supplement this deficiency of power with their tongue, which is very long and supple, and is in many kinds so divided that it can touch more than one object at a time; the habit of reconnoitring objects in front of them with their tongues has even led to their being able to pass it through the end of their nostrils without being obliged to open their jaws.[307]

"Herbivorous mammals, such as the elephant, rhinoceros, ox, buffalo, horse, &c., owe their great size to their habit of daily distending themselves with food and taking comparatively little exercise. They employ their feet for standing, walking, or running, but not for climbing trees. Hence the thick horn which covers their toes. These toes have become useless to them, and are now in many cases rudimentary only. Some pachyderms have five toes covered with horn; some four, some three. The ruminants, which appear to be the earliest mammals that confined themselves to a life upon the ground, have but two hooves, while the horse has only one.[308]

"Some herbivorous animals, especially among the ruminants, have been incessantly preyed upon by carnivorous animals, against which their only refuge is in flight. Necessity has therefore developed the light and active limbs of antelopes, gazelles, &c. Ruminants, only using their jaws to graze with, have but little power in them, and therefore generally fight with their heads. The males fight frequently with one another, and their desires prompt an access of fluids to the parts of their heads with which they fight; thus the horns and bosses have arisen with which the heads of most of these animals are armed.[309] The giraffe owes its long neck to its continued habit of browsing upon trees, whence also the great length of its fore legs as compared with its hinder ones. Carnivorous animals, in like manner, have had their organs modified in correlation with their desires and habits. Some climb, some scratch in order to burrow in the earth, some tear their prey; they therefore have need of toes, and we find their toes separated and armed with claws. Some of them are great hunters, and also plunge their claws deeply into the bodies of their victims, trying to tear out the part on which they have seized; this habit has developed a size and curvature of claw which would impede them greatly in travelling over stony ground; they have therefore been obliged to make efforts to draw back their too projecting claws, and so, little by little, has arisen the peculiar sheath into which cats, tigers, lions, &c., withdraw their claws when they no longer wish to use them.[310]

"We see then that the long-sustained and habitual exercise of any part of a living organism, in consequence of the necessities engendered by its environment, develops such part, and gives it a form which it would never have attained if the exercise had not become an habitual action. All known animals furnish us with examples of this.[311] If anyone maintains that the especially powerful development of any organ has had nothing to do with its habitual use--that use has added nothing, and disuse detracted nothing from its efficiency, but that the organ has always been as we now see it from the creation of the particular species onwards--I would ask why cannot our domesticated ducks fly like wild ducks? I would also quote a multitude of examples of the effects of use and disuse upon our own organs, effects which, if the use and disuse were constant for many generations, would become much more marked.

"A great number of facts show, as will be more fully insisted on, that when its will prompts an animal to this or that action, the organs which are to execute it receive an excess of nervous fluid, and this is the determinant cause of the movements necessary for the required action. Modifications acquired in this way eventually become permanent in the breed that has acquired them, and are transmitted to offspring, without the offspring's having itself gone through the processes of acquisition which were necessary in the case of the ancestor.[312] Frequent crosses, however, with unmodified individuals, destroy the effect produced. It is only owing to the isolation of the races of man through geographical and other causes, that man himself presents so many varieties, each with a distinctive character.

"A review of all existing classes, orders, genera, and species would show that their structure, organs, and faculties, are in all cases solely attributable to the surroundings to which each creature has been subjected by nature, and to the habits which individuals have been compelled to contract; and that they are not at all the result of a form originally bestowed, which has imposed certain habits upon the creature.[313]

"It is unnecessary to multiply instances; the fact is simply this, that all animals have certain habits, and that their organization is always in perfect harmony with these habits.[314] The conclusion hitherto accepted is that the Author of Nature, when he created animals, foresaw all the possible circumstances in which they would be placed, and gave an unchanging organism to each creature, in accordance with its future destiny. The conclusion, on the other hand, here maintained is that nature has evolved all existing forms of life successively, beginning with the simplest organisms and gradually proceeding to those which are more complete. Forms of life have spread themselves throughout all the habitable parts of the earth, and each species has received its habits and corresponding modification of organs, from the influence of the surroundings in which it found itself placed.[315]

"The first conclusion supposes an unvarying organism and unvarying conditions. The second, which is my theory (*la mienne propre*), supposes that each animal is capable of modifications which in the course of generations amount to a wide divergence of type.

"If a single animal can be shown to have varied considerably under domestication, the first conclusion is proved to be inadmissible, and the second to be in conformity with the laws of nature."

This is a milder version of Buffon's conclusion (see *ante*, pp. 90, 91). It is a little grating to read the words "la mienne propre," and to recall no mention of Buffon in the 'Philosophie Zoologique.'

"Animal forms then are the result of conditions of life and of the habits engendered thereby. With new forms new faculties are developed, and thus nature has little by little evolved the existing differentiations of animal and vegetable life."[316]

Lamarck makes no exception in man's favour to the rule of descent with modification. He supposes that a race of quadrumanous apes gradually acquired the upright position in walking, with a corresponding modification of the feet and facial angle. Such a race having become master of all the other animals, spread itself over all parts of the world that suited it. It hunted out the other higher races which were in a condition to dispute with it for enjoyment of the world's productions, and drove them to take refuge in such places as it did not desire to occupy. It checked the increase of the races nearest itself, and kept them exiled in woods and desert places, so that their further development was arrested, while itself, able to spread in all directions, to multiply without opposition, and to lead a social life, it developed new requirements one after another, which urged it to industrial pursuits, and gradually perfected its capabilities. Eventually this pre-eminent race, having acquired absolute supremacy, came to be widely different from even the most perfect of the lower animals.

"Certain apes approach man more nearly than any

other animal approaches him; nevertheless, they are far inferior to him, both in bodily and mental capacity. Some of them frequently stand upright, but as they do not habitually maintain this attitude, their organization has not been sufficiently modified to prevent it from being irksome to them to stand for long together. They fall on all fours immediately at the approach of danger. This reveals their true origin.[317]

"But is the upright position altogether natural, even to man? He uses it in moving from place to place, but still standing is a fatiguing position, and one which can only be maintained for a limited time, and by the aid of muscular contraction. The vertebrate column does not pass through the axis of the head so as to maintain it in like equilibrium with other limbs. The head, chest, stomach, and intestines weigh almost entirely on the anterior part of the vertebrate column, and this column itself is placed obliquely, so that, as M. Richerand has observed, continual watchfulness and muscular exertion are necessary to avoid the falls towards which the weight and disposition of our parts are continually inclining us. 'Children,' he remarks, 'have a constant tendency to assume the position of quadrupeds.'"[318]

"Surely these facts should reveal man's origin as analogous to that of the other mammals, if his organization only be looked to. But the following consideration must be added. New wants, developed in societies which had become numerous, must have correspondingly multiplied the ideas of this dominant race, whose individuals must have therefore gradually felt the need of fuller communication with each other. Hence the necessity for increasing and varying the number of the signs suitable for mutual understanding. It is plain therefore that incessant efforts would be made in this direction.[319]

"The lower animals, though often social, have been kept in too great subjection for any such development of power. They continue, therefore, stationary as regards their wants and ideas, very few of which need be communicated from one individual to another. A few movements of the body, a few simple cries and whistles, or inflexions of voice, would suffice for their purpose. With the dominant race, on the other hand, the continued multiplication of ideas which it was desirable to communicate rapidly, would exhaust the power of pantomimic gesture and of all possible inflexions of the voice--therefore by a succession of efforts this race arrived at the utterance of articulate sounds. A few only would be at first made use of, and these would be supplemented by inflexions of the voice: presently they would increase in number, variety, and appropriateness, with the increase of needs and of the efforts made to speak. Habitual exercise would increase the power of the lips and tongue to articulate distinctly.

"The diversity of language is due to geographical distribution, with consequent greater or less isolation of certain races, and corruption of the signs originally agreed upon for each idea. Man's own wants, therefore, will have achieved the whole result. They will have given rise to endeavour, and habitual use will have developed the organs of articulation."[320]

How, let me ask again, is "the case of neuter insects" "demonstrative" against the "well-known" theory put forward in the foregoing chapter?

FOOTNOTES:

[208] 'Phil. Zool.,' tom. i., edited by M. Martins, 1873, pp. 25, 26.

[209] 'Phil. Zool.' tom. i. pp. 26, 27.

[210] Page 28.

[211] Pages 28-31.

[212] 'Phil. Zool.,' tom. i. pp. 34, 35.

[213] Page 42.

[214] Page 46.

[215] 'Phil. Zool.,' tom. i. p. 50.

[216] Pages 50, 51.

[217] 'Origin of Species,' p. 395, ed. 1876.

[218] 'Phil. Zool.,' tom. i. p. 61.

[219] 'Phil. Zool.,' tom. i. p. 62.

[220] Page 63.

[221] Page 64.

[222] Page 65.

[223] Page 67.

[224] Chap. iii.

[225] 'Phil. Zool.,' tom. i. p. 72.

[226] Pages 71-73.

[227] 'Phil. Zool.,' tom. i. p. 74, 75.

[228] 'Phil. Zool.,' tom. i. pp. 75-77.

[229] 'Origin of Species,' p. 104, ed. 1876.

[230] 'Phil. Zool.,' tom. i. p. 79.

[231] 'Phil. Zool.,' tom. i. pp. 79, 80.

[232] 'Phil. Zool.,' tom. i. p. 80.

[233] Page 80.

[234] Ed. 1876.

[235] 'Phil. Zool.,' tom. i. p. 81.

[236] 'Origin of Species,' p. 241.

[237] 'Phil. Zool.,' p. 82.

[238] 'Phil. Zool.,' tom. i. p. 83.

[239] Pages 349-351.

[240] Page 84.

[241] 'Phil. Zool.,' tom. i. p. 88.

[242] Page 90.

[243] 'Origin of Species,' p. 3.

[244] 'Phil. Zool.,' tom. i. p. 94.

[245] Pages 95-96.

[246] Page 97.

[247] Phil. Zool.,' tom. i. p. 98.

[248] 'Phil. Zool.,' tom. i. p. 111.

[249] 'Phil. Zool.,' tom. i. p. 112.

[250] See pp. 227 and 259 of this book.

[251] 'Phil. Zool.,' tom. i. p. 113.

[252] Page 113.

[253] 'Phil Zool.,' tom. i. p. 113.

[254] This passage is rather obscure. I give it therefore in the original:--

"Ainsi les naturalistes ayant remarqué que beaucoup d'espèces, certains genres, et même quelques familles paraissent dans une sorte d'isolement, quant à leurs caractères, plusieurs se sont imaginés que les êtres vivants, dans l'un ou l'autre règne, s'avoisinaient, ou s'éloignaient entre eux, relativement à leurs *rapports naturels*, dans une disposition semblable aux differents points d'une carte de géographie ou d'une mappemonde. Ils regardent les petites séries bien prononcées qu'on a nommées familles naturelles, comme devant être disposées entre elles de manière à former une réticulation. Cette idée qui a paru sublime à quelques modernes, est évidemment une erreur, et, sans doute, elle se dissipera dès qu'on aura des connaissances plus profondes et plus générales de l'organisation, et surtout lorsqu'on distinguera ce qui appartient à l'influence des lieux d'habitation et des habitudes contractées, de ce qui résulte des progrès plus ou moins avancés dans la composition ou le perfectionnement de l'organisation."--(p. 120).

[255] 'Origin of Species,' pp. 265, 266.

[256] 'Phil. Zool.,' tom. i. p. 121.

[257] 'Phil. Zool.,' tom. i. p. 122.

[258] 'Origin of Species,' pp. 122, 123.

[259] 'Phil. Zool.,' tom. i. p. 123.

[260] 'Phil. Zool.,' tom. i. p. 123.

[261] 'Origin of Species,' chap. xiv.

[262] 'Phil. Zool.,' tom. i. p. 123.

[263] 'Phil. Zool.,' tom. i. p. 140.

[264] Page 142.

[265] Page 143.

[266] 'Phil. Zool.,' tom. i. p. 143.

[267] Page 144.

[268] Ibid.

[269] 'Phil. Zool.,' tom. i. p. 145.

[270] Page 146.

[271] 'Phil. Zool.,' tom. i. p. 221.

[272] Page 222.

[273] 'Phil. Zool.,' tom. i. p. 223.

[274] Page 224.

[275] Page 223.

[276] Page 225.

[277] 'Phil. Zool.,' tom. i. p. 225.

[278] Page 226.

[279] 'Phil. Zool.,' tom. i. p. 228.

[280] See Buffon, 'Hist. Nat.,' tom. v. pp. 196, 197, and Supp. tom. v. pp. 250-253.

[281] 'Phil. Zool.,' tom. i. p. 229.

[282] 'Hist. Nat.,' tom. xi. p. 290.

[283] 'Phil. Zool.,' tom. i. p. 231.

[284] Page 231. See Dr. Darwin's note on *Trapa natans*, 'Botanic Garden,' part ii. canto 4, l. 204.

[285] 'Phil. Zool.,' tom. i. p. 232.

[286] Page 233. See Buffon on Climate, tom. ix., 'The Animals of the Old and New Worlds.'

[287] 'Origin of Species,' p. 233, ed. 1876.

[288] 'Phil. Zool.,' tom. i. p 234.

[289] Page 235.

[290] Page 236.

[291] 'Phil. Zool.,' tom. i. p. 237.

[292] Page 238.

[293] See *ante*, pp. 220-228.

[294] 'Phil. Zool.,' tom. i. p. 239.

[295] 'Phil. Zool.,' tom. i. p 240.

[296] Page 241.

[297] Page 245.

[298] 'Animals and Plants under Domestication,' vol. i. p. 467, &c.

[299] See frontispiece to Professor Mivart's 'Genesis of Species.'

[300] 'Phil. Zool.,' tom. i. p. 247.

[301] Page 248.

[302] 'Nat. Theol.,' vol. xii., end of § viii.

[303] 'Phil. Zool.,' tom. i. p. 249.

[304] 'Phil. Zool.,' tom. i. p. 250.

[305] Page 250.

[306] 'Phil. Zool.,' tom. i. p. 251.

[307] Page 252.

[308] 'Phil. Zool.,' tom. i. p. 253.

[309] Page 254.

[310] 'Phil. Zool.,' tom. i. p. 256.

[311] Page 257.

[312] 'Phil. Zool.,' tom. i. p. 259.

[313] Page 260.

[314] Page 263.

[315] 'Phil. Zool.,' tom. i. p. 263.

[316] Page 265.

[317] 'Phil. Zool.,' tom. i. p. 343.

[318] 'Phil. Zool.,' tom. i. p. 343.

[319] Page 346.

[320] 'Phil. Zool.,' tom. i. p. 347.

CHAPTER XVIII.

MR. PATRICK MATTHEW, MM. ÉTIENNE AND ISIDORE GEOFFROY ST. HILAIRE, AND MR. HERBERT SPENCER.

The same complaint must be made against Mr. Matthew's excellent survey of the theory of evolution, as against Dr. Erasmus Darwin's original exposition of the same theory, namely, that it is too short. It may be very true that brevity is the soul of wit, but the leaders of science will generally succeed in burking new-born wit, unless the brevity of its soul is found compatible with a body of some bulk.

Mr. Darwin writes thus concerning Mr. Matthew in the historical sketch to which I have already more than once referred.

"In 1831 Mr. Patrick Matthew published his work on 'Naval Timber and Arboriculture,' in which he gives precisely the same view on the origin of species as that (presently to be alluded to) propounded by Mr. Wallace and myself in the 'Linnean Journal,' and as that enlarged in the present volume. Unfortunately the view was given by Mr. Matthew very briefly, in scattered passages in an appendix to a work on a different subject, so that it remained unnoticed until Mr. Matthew himself drew attention to it in the 'Gardener's Chronicle' for April 7, 1860. The differences of Mr. Matthew's view from mine are not of much importance; he seems to consider that the world was nearly depopulated at successive periods, and then re-stocked, and he gives as an alternative, that new forms may be generated 'without the presence of any mould or germ of former aggregates.' I am not sure that I understand some passages; but it seems that he attributes much influence to the direct action of the conditions of life. He clearly saw, however, the full force of the principle of natural selection."[321]

Nothing could well be more misleading. If Mr. Matthew's view of the origin of species is "precisely the same as that" propounded by Mr. Darwin, it is hard to see how Mr. Darwin can call those of Lamarck and Dr. Erasmus Darwin "erroneous"; for Mr. Matthew's is nothing but an excellent and well-digested summary of the conclusions arrived at by these two writers and by Buffon. If, again, Mr. Darwin is correct in saying that Mr. Matthew "clearly saw the full force of the principle of natural selection," he condemns the view he has himself taken of it in his 'Origin of Species,' for Mr. Darwin has assigned a far more important and very different effect to the fact that the fittest commonly survive in the struggle for existence, than Mr. Matthew has done. Mr. Matthew sees a cause underlying all variations; he takes the most teleological or purposive view of organism that has been taken by any writer (not a theologian) except myself, while Mr. Darwin's view, if not the least teleological, is certainly nearly so, and his confession of inability to detect any general cause underlying variations, leaves, as will appear presently, less than common room for ambiguity. Here are Mr. Matthew's own words:--

"There is a law universal in nature, tending to render every reproductive being the best possibly suited to the condition that its kind, or that organized matter is susceptible of, and which appears intended to model the physical and mental or instinctive, powers to their highest perfection, and to continue them so. This law sustains the lion in his strength, the hare in her swiftness, and the fox in his wiles. As nature in all her modifications of life has a power of increase far beyond what is needed to supply the place of what falls by Time's decay, those individuals who possess not the requisite strength, swiftness, hardihood, or cunning, fall prematurely without reproducing--either a prey to their natural devourers, or sinking under disease, generally induced by want of nourishment, their place being occupied by the more perfect of their own kind, who are pressing on the means of existence.

"Throughout this volume, we have felt considerable inconvenience from the adopted dogmatical classification of plants, and have all along been floundering between species and variety, which certainly under culture soften into each other. A particular conformity, each after its own kind, when in a state of nature, termed species, no doubt exists to a considerable degree. This conformity has existed during the last forty centuries; geologists discover a like particular conformity--fossil species--through the deep deposition of each great epoch; but they also discover an almost complete difference to exist between the species or stamp of life of one epoch from that of every other. We are therefore led to admit either a repeated miraculous conception, or *a power of change under change of circumstances* to belong to living organized matter, or rather to the congeries of inferior life which appears to form superior." (By this I suppose Mr. Matthew to imply his assent to the theory, that our personality or individuality is but as it were "the consensus, or full flowing river of a vast number of subordinate individualities or personalities, each one of which is a living being with thoughts and wishes of its own.") "The derangements and changes in organized existence, induced by a change of circumstances from the interference of man, afford us proof of the plastic quality of superior life; and the likelihood that circumstances have been very different in the different epochs, though steady in each, tend strongly to heighten the probability of the latter theory.

"When we view the immense calcareous and bituminous formations, principally from the waters and atmosphere, and consider the oxidations and depositions which have taken place, either gradually or during some of the great convulsions, it appears at least probable that the liquid elements containing life have varied considerably at different times in composition and weight; that our atmosphere has contained a much greater proportion of carbonic acid or oxygen; and our waters, aided by excess of carbonic acid, and greater heat resulting from greater density of atmosphere, have contained a greater quantity of lime, and other mineral solutions. Is the inference, then, unphilosophic that living things which are proved to have *a circumstance-suiting power* (a very slight change of circumstance by culture inducing a corresponding change of character), may have gradually accommodated themselves to the variations of the elements containing them, and without new creation, have presented the diverging changeable phenomena of past and present organized existence?

"The destructive liquid currents before which the hardest mountains have been swept and comminuted into gravel, sand, and mud, which intervened between and divided these epochs, probably extending over the whole surface of the globe and destroying nearly all living things, must have reduced existence so much that an unoccupied field would be formed for new diverging ramifications of life, which from the connected sexual system of vegetables, and the natural instinct of animals to herd and combine with their own kind, would fall into specific groups--these remnants in the course of time moulding and accommodating their being anew to the change of circumstances, and to every possible means of subsistence--and the millions of ages of regularity which appear to have followed between the epochs, probably after this accommodation was completed, affording fossil deposit of regular specific character.

"In endeavouring to trace ... the principle of these changes of fashion which have taken place in the domiciles of life the following questions occur: Do they arise from admixture of species nearly allied producing intermediate species? Are they the diverging ramifications of the living principle under modification of circumstance? or have they resulted from the combined agency of both?

"*Is there only one living principle? Does organized existence, and perhaps all material existence, consist of one Proteus principle of life* capable of gradual circumstance-suited modifications and aggregations without bound, under the solvent or motion-giving principle of heat or light? There is more beauty and unity of design in this continual balancing of life to circumstance, and greater conformity to those dispositions of nature that are manifest to us, than in total destruction and new creation. It is improbable that much of this diversification is owing to commixture of species nearly allied; all change by this appears very limited and confined within the bounds of what is called species; the progeny of the same parents under great difference of circumstance, might in several generations even become distinct species, incapable of co-reproduction.

"The self-regulating adaptive disposition of organized life may, in part, be traced to the extreme fecundity of nature, who, as before stated, has in all the varieties of her offspring a prolific power much beyond (in many cases a thousand fold) what is necessary to fill up the vacancies caused by senile decay. As the field of existence is limited and preoccupied, it is only the hardier, more robust, better suited to circumstance individuals, who are able to struggle forward to maturity, these inhabiting only the situations to which they have *superior adaptation and greater power of occupancy than any other kind; the weaker and less circumstance-suited being prematurely destroyed*. This principle is in constant action; it regulates the colour, the figure, the capacities, and instincts; those individuals in each species whose colour and covering are best suited to concealment or protection from enemies, or defence from inclemencies and vicissitudes of climate, whose figure is best accommodated to health, strength, defence, and support; whose capacities and instincts can best regulate the physical energies to self-advantage according to circumstances--in such immense waste of primary and youthful life those only come forward to maturity from the strict ordeal by which nature tests their adaptation to her standard of perfection and fitness to continue their kind by reproduction.

"From the unremitting operation of this law acting in concert with the tendency which the progeny have to take the more particular qualities of the parents, together with the connected sexual system in vegetables and instinctive limitation to its own kind in animals, a considerable uniformity of figure, colour, and character is induced constituting species; the breed gradually acquiring the very best possible adaptation of these to its condition which it is susceptible of, and when alteration of circumstance occurs, thus changing in character to suit these, as far as its nature is susceptible of change.

"This circumstance-adaptive law operating upon the slight but continued natural disposition to sport in the progeny (seedling variety) *does not preclude the supposed influence which volition or sensation may have had over the configuration of the body*. To examine into the disposition to sport in the progeny, even when there is only one parent as in many vegetables, and to investigate how much variation is modified by the mind or nervous sensation of the parents, or of the living thing itself during its progress to maturity; how far it depends upon external circumstance, and how far on the will, irritability, and muscular exertion, is open to examination and experiment. In the first place, we ought to examine its dependency upon the preceding links of the particular chain of life, variety being often merely types or approximations of former parentage; thence the variation of the family as well as of the individual must be embraced by our experiments.

"This continuation of family type, not broken by casual particular aberration, is mental as well as corporeal, and is exemplified in many of the dispositions or instincts of particular races of men. *These innate or continuous ideas or habits seem proportionally greater in the insect tribes, and in those especially of shorter revolution; and forming an abiding memory, may resolve much of the enigma of instinct, and the foreknowledge which these tribes have of what is necessary to completing their round of life, reducing this to knowledge or impressions and habits acquired by a long experience.*

"This greater continuity of existence, or rather continuity of perceptions and impressions in insects, is highly probable; *it is even difficult in some to ascertain the particular steps when each individual commences*, under the different phases of egg, larva, pupa, or if much consciousness of individuality exists. The continuation of reproduction for several generations by the females alone in some of these tribes, *tends to the probability of the greater continuity of existence; and the subdivisions of life by cuttings (even in animal life), at any rate, must stagger the advocate of individuality.*

"Among the millions of specific varieties of living things which occupy the humid portions of the surface of our planet, as far back as can be traced, there does not appear, with the exception of man, to have been any particular engrossing race, but a pretty fair balance of power of occupancy--or rather most wonderful variation of circumstance parallel to the nature of every species, *as if circumstance and species had grown up together*. There are, indeed, several races which have threatened ascendancy in some particular regions; but it is man alone from whom any general imminent danger to the existence of his brethren is to be dreaded.

"As far back as history reaches, man had already had considerable influence, and had made encroachments upon his fellow denizens, probably occasioning the destruction of many species, and the production and continuation of a number of varieties, and even species, which he found more suited to supply his wants, but which from the infirmity of their condition--*not having undergone selection by the law of nature*, of which we have spoken--cannot maintain their ground without culture and protection.

"It is only however in the present age that man has begun to reap the fruits of his tedious education, and has proven how much 'knowledge is power.' He has now acquired a dominion over the material world, and a consequent power of increase, so as to render it probable that the whole surface of the earth may soon be overrun by this engrossing anomaly, to the annihilation of every wonderful and beautiful variety of animal existence which does not administer to his wants, principally as laboratories of preparation to befit cruder elemental matter for assimilation by his organs.

"The consequences are being now developed of our deplorable ignorance of, or inattention to, one of the most evident traits of natural history--that vegetables, as well as animals, are generally liable to an almost unlimited diversification, regulated by climate, soil, nourishment, and new commixture of already-formed varieties. In those with which man is most intimate, and where his agency in throwing them from their natural locality and disposition has brought out this power of diversification in stronger shades, it has been forced upon his notice, as in man himself, in the dog, horse, cow, sheep, poultry,--in the apple, pear, plum, gooseberry, potato, pea, which sport in infinite varieties, differing considerably in size, colour, taste, firmness of texture, period of growth, almost in every recognizable quality. In all these kinds man is influential in preventing deterioration, by careful selection of the largest or most valuable as breeders."[322]

Étienne and Isidore Geoffroy.

"Both Cuvier and Étienne Geoffroy," says Isidore Geoffroy, "had early perceived the philosophical importance of a question (evolution) which must be admitted as--with that of unity of composition--the greatest in natural history. We find them laying it down in the year 1795 in one of their joint 'Memoirs' (on the Orangs), in the very plainest terms, in the following question, 'Must we see,' they inquire, 'what we commonly call species, as the modified descendants of the same original form?'

"Both were at that time doubtful. Some years afterwards Cuvier not only answered this question in the negative, but declared, and pretended to prove, that the same forms have been perpetuated from the beginning of things. Lamarck, his antagonist *par excellence* on this point, maintained the contrary position with no less distinctness, showing that living beings are unceasingly variable with change of their surroundings, and giving with some boldness a zoological genesis in conformity with this doctrine.

"Geoffroy St. Hilaire had long pondered over this difficult subject. The doctrine which in his old age he so firmly defended, does not seem to have been conceived by him till after he had completed his 'Philosophie Anatomique,' and except through lectures delivered orally to the museum and the faculty, it was not published till 1828; nor again in the work then published do we find his theory in its neatest expression and fullest development."

Isidore Geoffroy St. Hilaire tells us in a note that the work referred to as first putting his father's views before the public in a printed form, was a report to the Academy of Sciences on a memoir by M. Roulin; but that before this report some indications of them are to be found in a paper on the Gavials, published in 1825. Their best rendering, however, and fullest development is in several memoirs, published in succession, between the years 1828 and 1837.

"This doctrine," he continues, "is diametrically opposed to that of Cuvier, and is not entirely the same as Lamarck's. Geoffroy St. Hilaire refutes the one, he restrains and corrects the other. Cuvier, according to him, sums up against the facts, while Lamarck goes further than they will bear him out. Essentially however on questions of this nature he is a follower of Lamarck, and took pleasure on several occasions in describing himself as the disciple of his illustrious *confrère*."[323]

I have been unable to detect any substantial difference of opinion between Geoffroy St. Hilaire and Lamarck, except that the first maintained that a line must be drawn somewhere--and did not draw it--while the latter said that no line could be drawn, and therefore drew none. Mr. Darwin is quite correct in saying that Geoffroy St. Hilaire "relied chiefly on the conditions of life, or the 'monde ambiant,' as the cause of change." But this is only Lamarck over again, for though Lamarck attributes variation directly to change of habits in the creature, he is almost wearisome in his insistence on the fact that the habit will not change, unless the conditions of life also do so. With both writers then it is change in the relative positions of the exterior circumstances, and of the organism, which results in variation, and finally in specific modification.

Here is another sketch of Étienne Geoffroy, also by his son Isidore.

In 1795, while Lamarck was still a believer in immutability, Étienne Geoffroy St. Hilaire "had ventured to say that species might well be 'degenerations from a single type,'" but, though he never lost sight of the question, he waited more than a quarter of a century before passing from meditation to action. "He at length put forward his opinion in 1825, he returned to it, but still briefly, in 1828 and 1829, and did not set himself to develop and establish it till the year 1831--the year following the memorable discussion in the Academy, on the unity of organic composition."[324]

"If," says his son, "he began by paying homage to his illustrious precursor, and by laying it down as a general axiom, that there is no such thing as fixity in nature, and especially in animated nature, he follows this adhesion to the general doctrine of variability by a dissent which goes to the very heart of the matter. And this dissent becomes deeper and deeper in his later works. Not only is Geoffroy St. Hilaire at pains to deny the unlimited extension of variability which is the foundation of the Lamarckian system, but he moreover and particularly declines to explain those degenerations which he admits as possible, by changes of action and habit on the part of the creature varying--Lamarck's favourite hypothesis, which he laboured to demonstrate without even succeeding in making it appear probable."[325]

Isidore Geoffroy then declares that his father, "though chronologically a follower of Lamarck, should be ranked philosophically as having continued the work of Buffon, to whom all his differences of opinion with Lamarck serve to bring him nearer."[326] If he had understood Buffon he would not have said so.

His conclusions are thus summed up:--"Geoffroy St. Hilaire maintains that species are variable if the environment varies in character; differences, then, more or less considerable according to the power of the modifying causes *may have* been produced in the course of time, and the living forms of to-day *may be* the descendants of more ancient forms."[327]

It is not easy to see that much weight should be attached to Geoffroy St. Hilaire's opinion. He seems to have been a person of hesitating temperament, under an impression that there was an opening just then through which a judicious trimmer might pass himself in among men of greater power. If his son has described his teaching correctly, it amounts practically to a *bonâ fide* endorsement of what Buffon can only be considered to have pretended to believe. The same objection that must be fatal to the view pretended by Buffon, is so in like manner to those put forward seriously of both the Geoffroys--for Isidore Geoffroy followed his father, but leant a little more openly towards Lamarck. He writes:--

"The characters of species are neither absolutely fixed, as has been maintained by some; nor yet, still more, indefinitely variable as according to others. They are fixed for each species as long as that species continues to reproduce itself in an unchanged environment; but they become modified if the environment changes."[328]

This is all that Lamarck himself would expect, as no

one could be more fully aware than M. Geoffroy, who, however, admits that degeneration may extend to generic differences.[329]

I have been unable to find in M. Isidore Geoffroy's work anything like a refutation of Lamarck's contention that the modifications in animals and plants are due to the needs and wishes of the animals and plants themselves; on the contrary, to some extent he countenances this view himself, for he says, "hence arise notable differences of habitation and climate, and these in their turn induce secondary differences in diet *and even in habits*."[330] From which it must follow, though I cannot find it said expressly, that the author attributes modification in some measure to changed habits, and therefore to the changed desires from which the change of habits has arisen; but in the main he appears to refer modification to the direct action of a changed environment.

Mr. Herbert Spencer.

"Those who cavalierly reject the theory of Lamarck and his followers as not adequately supported by facts," wrote Mr. Herbert Spencer,[331] "seem quite to forget that their own theory is supported by no facts at all"--inasmuch as no one pretends to have seen an act of direct creation. Mr. Spencer points out that, according to the best authorities, there are some 320,000 species of plants now existing, and about 2,000,000 species of animals, including insects, and that if the extinct forms which have successively appeared and disappeared be added to these, there cannot have existed in all less than some ten million species. "Which," asks Mr. Spencer, "is the most rational theory about these ten millions of species? Is it most likely that there have been ten millions of special creations? or, is it most likely that by continual modification *due to change of circumstances*, ten millions of varieties may have been produced as varieties are being produced still?"

"Even could the supporters of the development hypothesis merely show that the production of species by the process of modification is conceivable, they would be in a better position than their opponents. But they can do much more than this; they can show that the process of modification has effected and is effecting great changes in all organisms, subject to modifying influences ... they can show that any existing species--animal or vegetable--when placed under conditions different from its previous ones, *immediately begins to undergo certain changes of structure* fitting it for the new conditions. They can show that in successive generations these changes continue until ultimately the new conditions become the natural ones. They can show that in cultivated plants and domesticated animals, and in the several races of men, these changes have uniformly taken place. They can show that the degrees of difference, so produced, are often, as in dogs, greater than those on which distinctions of species are in other cases founded. They can show that it is a matter of dispute whether some of these modified forms *are* varieties or modified species. They can show too that the changes daily taking place in ourselves; the facility that attends long practice, and the loss of aptitude that begins when practice ceases; the strengthening of passions habitually gratified, and the weakening of those habitually curbed; the development of every faculty, bodily, moral or intellectual, according to the use made of it, are all explicable on this same principle. And thus they can show that throughout all organic nature there *is* at work a modifying influence of the kind they assign as the cause of these specific differences, an influence which, though slow in its action, does in time, if the circumstances demand it, produce marked changes; an influence which, to all appearance, would produce in the millions of years, and under the great varieties of condition which geological records imply, any amount of change."

This leaves nothing to be desired. It is Buffon, Dr. Darwin, and Lamarck, well expressed. Those were the days before "Natural Selection" had been discharged into the waters of the evolution controversy, like the secretion of a cuttle fish. Changed circumstances immediately induce changed habits, and hence a changed use of some organs, and disuse of others: as a consequence of this, organs and instincts become changed, "and these changes continue in successive generations, until ultimately the new conditions become the natural ones." This is the whole theory of "development," "evolution," or "descent with modification." Volumes may be written to adduce the details which warrant us in accepting it, and to explain the causes which have brought it about, but I fail to see how anything essential can be added to the theory itself, which is here so well supported by Mr. Spencer, and which is exactly as Lamarck left it. All that remains is to have a clear conception of the oneness of personality between parents and offspring, of the eternity, and latency, of memory, and of the unconsciousness with which habitual actions are repeated, which last point, indeed, Mr. Spencer has himself touched upon.

Mr. Spencer continues--"That by any series of changes a zoophyte should ever become a mammal, seems to those who are not familiar with zoology, and who have not seen how clear becomes the relationship between the simplest and the most complex forms, when all intermediate forms are examined, a very grotesque notion ... they never realize the fact that by small increments of modification, any amount of modification may in time be generated. That surprise which they feel on finding one whom they last saw as a boy, grown into a man, becomes incredulity when the degree of change is greater. Nevertheless, abundant instances are at hand of the mode in which we may pass to the most diverse forms by insensible gradations."

Nothing can be more satisfactory and straightforward. I will make one more quotation from this excellent article:--

"But the blindness of those who think it absurd to suppose that complex organic forms may have arisen by successive modifications out of simple ones, becomes astonishing when we remember that complex organic forms are daily being thus produced. A tree differs from a seed immeasurably in every respect--in bulk, in structure, in colour, in form, in specific gravity, in chemical composition--

differs so greatly that no visible resemblance of any kind can be pointed out between them. Yet is the one changed in the course of a few years into the other--changed so gradually that at no moment can it be said, 'Now the seed ceases to be, and the tree exists.' What can be more widely contrasted than a newly-born child, and the small, semi-transparent gelatinous spherule constituting the human ovum? The infant is so complex in structure that a cyclopædia is needed to describe its constituent parts. The germinal vesicle is so simple, that a line will contain all that can be said of it. Nevertheless, a few months suffices to develop the one out of the other, and that too by a series of modifications so small, that were the embryo examined at successive minutes, not even a microscope would disclose any sensible changes. That the uneducated and ill-educated should think the hypothesis that all races of beings, man inclusive, may in process of time have been evolved from the simplest monad a ludicrous one is not to be wondered at. But for the physiologist, who knows that every individual being *is* so evolved--who knows further that in their earliest condition the germs of all plants and animals whatsoever are so similar, 'that there is no appreciable distinction among them which would enable it to be determined whether a particular molecule is the germ of a conferva or of an oak, of a zoophyte or of a man'[332]--for him to make a difficulty of the matter is inexcusable. Surely, if a single structureless cell may, when subjected to certain influences, become a man in the space of twenty years, there is nothing absurd in the hypothesis that under certain other influences a cell may, in the course of millions of years, give origin to the human race. The two processes are generically the same, and differ only in length and complexity."

The very important extract from Professor Hering's lecture should perhaps have been placed here. The reader will, however, find it on page 199.

FOOTNOTES:

[321] 'Origin of Species,' Hist. Sketch, p. xvi.

[322] See 'Naval Timber and Arboriculture,' by Patrick Matthew, published by Adam and C. Black, Edinburgh, and Longmans and Co., London, 1831, pp. 364, 365, 381-388, and also 106-108, 'Gardeners' Chronicle,' April 7, 1860.

[323] 'Vie et Doctrine Scientifique de Geoffroy Étienne St. Hilaire,' Paris, Strasbourg, 1847, pp. 344-346.

[324] 'Hist. Nat. Gén.,' tom. ii. 413.

[325] 'Hist. Nat. Gén.,' tom. ii. p. 415.

[326] Ibid.

[327] Ibid. p. 421.

[328] 'Hist. Nat. Gén.,' vol. ii. p. 431, 1859.

[329] 'Origin of Species,' Hist. Sketch, p. xix.

[330] 'Hist. Nat. Gén.,' vol. ii. p. 432.

[331] See 'The Leader,' March 20, 1852, "The Haythorne Papers."

[332] Carpenter's 'Principles of Physiology', 3rd ed., p. 867.

CHAPTER XIX.

MAIN POINTS OF AGREEMENT AND OF DIFFERENCE BETWEEN THE OLD AND NEW THEORIES OF EVOLUTION.

Having put before the reader with some fulness the theories of the three writers to whom we owe the older or teleological view of evolution, I will now compare that view more closely with the theory of Mr. Darwin and Mr. Wallace, to whom, in spite of my profound difference of opinion with them on the subject of natural selection, I admit with pleasure that I am under deep obligation. For the sake of brevity, I shall take Lamarck as the exponent of the older view, and Mr. Darwin as that of the one now generally accepted.

We have seen, that up to a certain point there is very little difference between Lamarck and Mr. Darwin. Lamarck maintains that animals and plants vary: so does Mr. Darwin. Lamarck maintains that variations having once arisen have a tendency to be transmitted to offspring and accumulated: so does

Mr. Darwin. Lamarck maintains that the accumulation of variations, so small, each one of them, that it cannot be, or is not noticed, nevertheless will lead in the course of that almost infinite time during which life has existed upon earth, to very wide differences in form, structure, and instincts: so does Mr. Darwin. Finally, Lamarck declares that all, or nearly all, the differences which we observe between various kinds of animals and plants are due to this exceedingly gradual and imperceptible accumulation, during many successive generations, of variations each one of which was in the outset small: so does Mr. Darwin. But in the above we have a complete statement of the fact of evolution, or descent with modification--wanting nothing, but entire, and incapable of being added to except in detail, and by way of explanation of the causes which have brought the fact about. As regards the general conclusion arrived at, therefore, I am unable to detect any difference of opinion between Lamarck and Mr. Darwin. They are both bent on establishing the theory of evolution in its widest extent.

The late Sir Charles Lyell, in his 'Principles of Geology,' bears me out here. In a note to his *résumé* of the part of the 'Philosophie Zoologique' which bears upon evolution, he writes:--

"I have reprinted in this chapter word for word my abstract of Lamarck's doctrine of transmutation, as drawn up by me in 1832 in the first edition of the 'Principles of Geology.'[333] I have thought it right to do this in justice to Lamarck, in order to show how nearly the opinions taught by him at the commencement of this century resembled those now in vogue amongst a large body of naturalists respecting the infinite variability of species, and the progressive development in past time of the organic world. The reader must bear in mind that when I made this analysis of the 'Philosophie Zoologique' in 1832, I was altogether opposed to the doctrine that the animals and plants now living were the lineal descendants of distinct species, only known to us in a fossil state, and ... so far from exaggerating, I did not do justice to the arguments originally adduced by Lamarck and Geoffroy St. Hilaire, especially those founded on the occurrence of rudimentary organs. There is therefore no room for suspicion that my account of the Lamarckian hypothesis, written by me thirty-five years ago, derived any colouring from my own views tending to bring it more into harmony with the theory since propounded by Darwin."[334] So little difference did Sir Charles Lyell discover between the views of Lamarck and those of his successors.

With the identity, however, of the main proposition which, both Lamarck and Mr. Darwin alike endeavour to establish, the points of agreement between the two writers come to an end. Lamarck's great aim was to discover the cause of those variations whose accumulation results in specific, and finally in generic, differences. Not content with establishing the fact of descent with modification, he, like his predecessors, wishes to explain how it was that the fact came about. He finds its explanation in changed surroundings--that is to say, in changed conditions of existence--as the indirect cause, and in the varying needs arising from these

changed conditions as the direct cause.

According to Lamarck, there is a broad principle which underlies variation generally, and this principle is the power which all living beings possess of slightly varying their actions in accordance with varying needs, coupled with the fact observable throughout nature that use develops, and disuse enfeebles an organ, and that the effects, whether of use or disuse, become hereditary after many generations.

This resolves itself into the effect of the mutual interaction of mind on body and of body on mind. Thus he writes:--

"The physical and the mental are to start with undoubtedly one and the same thing; this fact is most easily made apparent through study of the organization of the various orders of known animals. From the common source there proceeded certain effects, and these effects, in the outset hardly separated, have in the course of time become so perfectly distinct, that when looked at in their extremest development they appear to have little or nothing in common.

"The effect of the body upon the mind has been already sufficiently recognized; not so that of the mind upon the body itself. The two, one in the outset though they were, interact upon each other more and more the more they present the appearance of having become widely sundered, and it can be shown that each is continually modifying the other and causing it to vary."[335]

And again, later:--

"I shall show that the habits by which we now recognize any creature are due to the environment (*circonstances*) under which it has for a long while existed, *and that these habits have had such an influence upon the structure of each individual of the species as to have at length*" (that is to say, through many successive slight variations, each due to habit engendered by the wishes of the animal itself), "modified this structure and adapted it to the habits contracted."[336]

These quotations must suffice, for the reader has already had Lamarck's argument sufficiently put before him.

Variation, and consequently modification, are, according to Lamarck, the outward and visible signs of the impressions made upon animals and plants in the course of their long and varied history, each organ chronicling a time during which such and such thoughts and actions dominated the creature, and specific changes being the effect of certain long-continued wishes upon the body, and of certain changed surroundings upon the wishes. Plants and animals are living forms of faith, or faiths of form, whichever the reader pleases.

Mr. Darwin, on the other hand, repeatedly avows ignorance, and profound ignorance, concerning the causes of those variations which, or nothing, must be the fountain-heads of species. Thus he writes of "the complex and *little known* laws of variation."[337] "There is also *some probability* in the view propounded by Andrew Knight, that variability *may be partly* connected with excess of food."[338] "Many laws regulate variation, *some few of which* can be *dimly seen*."[339] "The results of the *unknown*, or *but dimly understood*, laws of variation are infinitely complex and diversified."[340] "We are *profoundly ignorant* of the cause of each slight variation or individual difference."[341] "We are *far too ignorant* to speculate on the relative importance of the several known and unknown causes of variation."[342] He admits, indeed, the effects of use and disuse to have been important, but how important we have no means of knowing; he also attributes considerable effect to the action of changed conditions of life--but how considerable again we know not; nevertheless, he sees no great principle underlying the variations generally, and tending to make them appear for a length of time together in any definite direction advantageous to the creature itself, but either expressly, as at times, or by implication, as throughout his works, ascribes them to accident or chance.

In other words, he admits his ignorance concerning them, and dwells only on the accumulation of variations the appearance of which for any length of time in any given direction he leaves unaccounted for.

Lamarck, again, having established his principle that sense of need is the main direct cause of variation, and having also established that the variations thus engendered are inherited, so that divergences accumulate and result in species and genera, is comparatively indifferent to further details. His work is avowedly an outline. Nevertheless, we have seen that he was quite alive to the effects of the geometrical ratio of increase, and of the struggle for existence which thence inevitably follows.

Mr. Darwin, on the other hand, comparatively indifferent to, or at any rate silent concerning the causes of those variations which appeared so all-important to Lamarck, inasmuch as they are the raindrops which unite to form the full stream of modification, goes into very full detail upon natural selection, or the survival of the fittest, and maintains it to have been "the most important but not the exclusive means of modification."[343]

It will be readily seen that, according to Lamarck, the variations which when accumulated amount to specific and generic differences, will have been due to causes which have been mainly of the same kind for long periods together. Conditions of life change for the most part slowly, steadily, and in a set direction; as in the direction of steady, gradual increase or decrease of cold or moisture; of the steady, gradual increase of such and such an enemy, or decrease of such and such a kind of food; of the gradual upheaval or submergence of such and such a continent, and consequent drying up or encroachment of such and such a sea, and so forth. The thoughts of the creature varying will thus have been turned mainly in one direction for long together; and hence the consequent modifications will also be mainly in fixed and definite directions for many successive generations; as in the direction of a warmer or cooler covering; of a better means of defence or of attack in relation to such and such another species; of a longer neck and longer legs, or of whatever other modification the gradually changing circumstances may be rendering expedient. It is easy to understand the accumulation of slight successive modifications which thus make their appearance in given organs and in a set direction.

With Mr. Darwin, on the contrary, the variations being accidental, and due to no special and uniform cause, will not appear for any length of time in any given direction, nor in any given organ, but will be just as liable to appear in one organ as in another, and may be in one generation in one direction, and in another in another.

In confirmation of the above, and in illustration of the important consequences that will follow according as we adopt the old or the more recent theory, I would quote the following from Mr. Mivart's 'Genesis of Species.'

Shortly before maintaining that two similar structures have often been developed independently of one another, Mr. Mivart points out that if we are dependent upon indefinite variations only, as provided for us by Mr. Darwin, this would be "so improbable as to be practically impossible."[344] The number of possible variations being indefinitely great, "it is therefore an indefinitely great number to one against a similar series of variations occurring and being similarly preserved in any two independent instances." It will be felt (as

Mr. Mivart presently insists) that this objection does not apply to a system which maintains that in case an animal feels any given want it will gradually develop the structure which shall meet the want--that is to say, if the want be not so great and so sudden as to extinguish the creature to which it has become a necessity. For if there be such a power of self-adaptation as thus supposed, two or more very widely different animals feeling the same kind of want might easily adopt similar means to gratify it, and hence develop eventually a substantially similar structure; just as two men, without any kind of concert, have often hit upon like means of compassing the same ends. Mr. Spencer's theory--so Mr. Mivart tells us--and certainly that of Lamarck, whose disciple Mr. Spencer would appear to be,[345] admits "a certain peculiar, but limited power of response and adaptation in each animal and plant"--to the conditions of their existence. "Such theories," says Mr. Mivart, "have not to contend against the difficulty proposed, and it has been urged that even very complex extremely similar structures have again and again been developed quite independently one of the other, and this because the process has taken place not by merely haphazard, indefinite variations in all directions, but by the concurrence of some other internal natural law or laws co-operating with external influences and with Natural Selection in the evolution of organic forms.

"*It must never be forgotten that to admit any such constant operation of any such unknown natural cause is to deny the purely Darwinian theory which relies upon the survival of the fittest by means of minute fortuitous indefinite variations.*

"Among many other obligations which the author has to acknowledge to Professor Huxley, are the pointing out of this very difficulty, and the calling his attention to the striking resemblance between certain teeth of the dog, and of the thylacine, as one instance, and certain ornithic peculiarities of pterodactyles as another."[346]

In brief then, changed distribution of use and disuse in consequence of changed conditions of the environment is with Lamarck the main cause of modification. According to Mr. Darwin natural selection, or the survival of favourable but accidental variations, is the most important means of modification. In a word, with Lamarck the variations are definite; with Mr. Darwin indefinite.

FOOTNOTES:

[333] Vol. ii. chap. i.

[334] Vol. ii. chap, xxxiv., ed. 1872.

[335] 'Philosophie Zoologique,' ed. M. Martins, Paris, Lyons, 1873, tom. i. p. 24.

[336] 'Philosophie Zoologique,' tom. i. p. 72.

[337] 'Origin of Species,' p. 3.

[338] Ibid. p. 5.

[339] 'Origin of Species,' p. 8.

[340] Ibid. p. 9.

[341] Ibid. p. 158.

[342] Ibid. p. 159.

[343] 'Origin of Species,' p. 4.

[344] 'Genesis of Species,' p. 74, 1871.

[345] See *ante*, p. 330, line 1 after heading.

[346] 'Genesis of Species,' p. 76, ed. 1871.

CHAPTER XX.

NATURAL SELECTION CONSIDERED AS A MEANS OF MODIFICATION. THE CONFUSION WHICH THIS EXPRESSION OCCASIONS.

When Mr. Darwin says that natural selection is the most important "means" of modification, I am not sure that I understand what he wishes to imply by the word "means." I do not see how the fact that those animals which are best fitted for the conditions of their existence commonly survive in the struggle for life, can be called in any special sense a "means" of modification.

"Means" is a dangerous word; it slips too easily into

"cause." We have seen Mr. Darwin himself say that Buffon did not enter on "the *causes or means*"[347] of modification, as though these two words were synonymous, or nearly so. Nevertheless, the use of the word "means" here enables Mr. Darwin to speak of Natural Selection as if it were an active cause (which he constantly does), and yet to avoid expressly maintaining that it is a cause of modification. This, indeed, he has not done in express terms, but he does it by implication when he writes, "Natural Selection *might be most effective in giving* the proper colour to each kind of grouse, and in *keeping* that colour when once acquired." Such language, says the late Mr. G. H. Lewes, "is misleading;" it makes "selection an agent."[348]

It is plain that natural selection cannot be considered a cause of variation; and if not of variation, which is as the rain drop, then not of specific and generic modification, which are as the river; for the variations must make their appearance before they can be selected. Suppose that it is an advantage to a horse to have an especially hard and broad hoof, then a horse born with such a hoof will indeed probably survive in the struggle for existence, but he was not born with the larger and harder hoof *because of his subsequently surviving*. He survived because he was born fit--not, he was born fit because he survived. The variation must arise first and be preserved afterwards.

Mr. Darwin therefore is in the following dilemma. If he does not treat natural selection as a cause of variation, the 'Origin of Species' will turn out to have no *raison d'être*. It will have professed to have explained to us the manner in which species has originated, but it will have left us in the dark concerning the origin of those variations which, when added together, amount to specific and generic differences. Thus, as I said in 'Life and Habit,' Mr. Darwin will have made us think we know the whole road, in spite of his having almost ostentatiously blindfolded us at every step in the journey. The 'Origin of Species' would thus prove to be no less a piece of intellectual sleight-of-hand than Paley's 'Natural Theology.'

If, on the other hand, Mr. Darwin maintains natural selection to be a cause of variation, this comes to saying that when an animal has varied in an advantageous direction, the fact of its subsequently surviving in the struggle for existence is the cause of its having varied in the advantageous direction--or more simply still--that the fact of its having varied is the cause of its having varied.

And this is what we have already seen Mr. Darwin actually to say, in a passage quoted near the beginning of this present book. When writing of the eye he says, "Variation will cause the slight alterations;"[349] but the "slight alterations" *are* the variations; so that Mr. Darwin's words come to this--that "variation will cause the variations."

There does not seem any better way out of this dilemma than that which Mr. Darwin has adopted--namely, to hold out natural selection as "a means" of modification, and thenceforward to treat it as an efficient cause; but at the same time to protest again and again that it is not a cause. Accordingly he writes that "Natural Selection *acts only by the preservation and accumulation* of small inherited modifications,"[350]--that is to say, it has had no share in inducing or causing these modifications. Again, "What applies to one animal will apply throughout all time to all animals--*that is, if they vary, for otherwise natural selection can effect nothing*"[351]; and again, "for natural selection only *takes advantage of such variations as arise*"[352]--the variations themselves arising, as we have just seen, from variation.

Nothing, then, can be clearer from these passages than that natural selection is not a cause of modification; while, on thc other hand, nothing can be clearer, from a large number of such passages, as, for instance, "natural selection may be *effective* in *giving* and *keeping* colour,"[353] than that natural selection is an efficient cause; and in spite of its being expressly declared to be only a "means" of modification, it will be accepted as cause by the great majority of readers.

Mr. Darwin explains this apparent inconsistency thus:--He maintains that though the advantageous modification itself is fortuitous, or without known cause or principle underlying it, yet its becoming the predominant form of the species in which it appears is due to the fact that those animals which have been advantageously modified commonly survive in times of difficulty, while the unmodified individuals perish: offspring therefore is more frequently left by the favourably modified animal, and thus little by little the whole species will come to inherit the modification. Hence the survival of the fittest becomes a means of modification, though it is no cause of variation.

It will appear more clearly later on how much this amounts to. I will for the present content myself with the following quotation from the late Mr. G. H. Lewes in reference to it. Mr. Lewes writes:--

"Mr. Darwin seems to imply that the external conditions which cause a variation are to be distinguished from the conditions which accumulate and perfect such variation, that is to say, he implies a radical difference between the process of variation and the process of selection. This I have already said does not seem to me acceptable; the selection I conceive to be simply the variation which has survived."[354]

Certainly those animals and plants which are best fitted for their environment, or, as Lamarck calls it, "*circonstances*"--those animals, in fact, which are best fitted to comply with the conditions of their existence--are most likely to survive and transmit their especial fitness. No one would admit this more readily than Lamarck. This is no theory; it is a commonly observed fact in nature which no one will dispute, but it is not more "a means of modification" than many other commonly observed facts concerning animals.

Why is "the survival of the fittest" more a means of modification than, we will say, the fact that animals live at all, or that they live in successive generations, being born, continuing their species, and dying, instead of living on for ever as one single animal in the common acceptation of the term; or than that they eat and drink?

The heat whereby the water is heated, the water which is turned into steam, the piston on which the steam acts, the driving wheel, &c., &c., are all one as much as another a means whereby a train is made to go from one place to another; it is impossible to say that any one of them is the main means. So (*mutatis mutandis*) with modification. There is no reason therefore why "the survival of the fittest" should claim to be an especial "means of modification" rather than any other necessary adjunct of animal or vegetable life.

I find that the late Mr. G. H. Lewes has insisted on this objection in his 'Physical Basis of Mind.' I observe, also, that in the very passage in which he does so, Mr. Lewes appears to have been misled by Mr. Darwin's use of that dangerous word "means," and, at the same time, by his frequent treatment of natural selection as though it were an active cause; so that Mr. Lewes supposes Mr. Darwin to have fallen into the very error of which, as I have above shown, he is evidently struggling to keep clear--namely, that of maintaining natural selection to be a "cause" of variation. Mr. Lewes then continues:--

"He [Mr. Darwin] separates Natural Selection from all the primary causes of variation either internal or external--either as results of the laws of growth, of the correlations of variation, of use and disuse, &c., and limits it to the slow accumulation of such variations as are profitable in the struggle with competitors. And for his purpose this separation is necessary. But biological philosophy must, I think, regard the distinction as artificial, *referring only to one of the great factors in the production of species*."[355]

The fact that one in a brood or litter is born fitter for the conditions of its existence than its brothers and sisters, and, again, the causes that have led to this one's having been born fitter--which last is what the older evolutionists justly dwelt upon as the most interesting consideration in connection with the whole subject--are more noteworthy factors of modification than the factor that an animal, if born fitter for its conditions, will commonly survive longer in the struggle for existence. If the first of these can be explained in such a manner as to be accepted as true, or highly probable, we have a substantial gain to our knowledge. The second is little--if at all--better than a truism. Granted, if it were not generally the case that those forms are most likely to survive which are best fitted for the conditions of their existence, no adaptation of form to conditions of existence could ever have come about. "The survival of the fittest" therefore, or, perhaps better, "the fertility of the fittest," is thus a *sine quâ non* for modification. But, as we have just insisted, this does not render "the fertility of the fittest" an especial "means of modification," rather than any other *sine quâ non* for modification.

But, to look at the matter in another light. Mr. Darwin maintains natural selection to be "the most important but not the exclusive means of modification."

For "natural selection" substitute the words "survival of the fittest," which we may do with Mr. Darwin's own consent abundantly given.

To the words "survival of the fittest" add what is elided, but what is, nevertheless, unquestionably as much implied as though it were said openly whenever these words are used, and without which "fittest" has no force--I mean, "for the conditions of their existence."

We thus find that when Mr. Darwin says that natural selection is the most important, but not exclusive means of modification, he means that the survival in the struggle for existence of those creatures which are best fitted to comply with the conditions of their existence is the most important, but not exclusive means whereby the descendants of a creature, we will say, A, have become modified, so as to be now represented by a creature, we will say, B.

But the word "*circonstances*," so frequently used by Lamarck for the conditions of an animal's existence, contains, by implication, the idea of animals *which shall exist or not according as they fulfil those conditions or fail to fulfil them*. Conditions of existence are conditions which something capable of existing must fulfil if it would exist at all, and nothing is a condition of an animal's existence which that animal need not comply with and may yet continue to exist. Again, the words "animals" and "plants" comprehend the ideas of "fit," "fitter," and "fittest," "unfit," "unfitter," and "unfittest" for certain conditions, for we know of no animals or plants in which we do not observe degrees of fitness or unfitness for their "*circonstances*" or environment, or conditions of existence.

The use, therefore, of the term "conditions of existence" is sufficient to show that the person using it intends to imply that those animals and plants will live longest (or survive) and thrive best which are best able to fulfil those conditions. Hence it implies neither more nor less than what is implied by the words "struggle for existence, with consequent survival of the fittest"--that is to say, if we hold the complying with any condition of life to which difficulty is attached to be part of "the struggle" for life, and this we should certainly do.

The words "conditions of existence" may, then, be used instead of the "struggle for existence with consequent survival of the fittest," for as they cannot imply any less than the "struggle, &c.," when they are set out in full, and without suppression, so neither do they imply more; for nothing is a condition of existence, in so far as its power of effecting the modification of any animal is concerned, which does not also involve more or less difficulty or struggle; for if there is no difficulty or struggle there will be nothing to bring about change of habit, and hence of structure. This identity of meaning may be also seen if we call to mind that the conditions of existence can be only a synonym for "the conditions of continuing to live," and "the conditions of continuing to live" a synonym for "the conditions of continuing to live a longer time," and "the conditions of continuing to live a longer time," for "the conditions of survival," and "the conditions of survival," for "the survival of the fittest," inasmuch as the being fittest is the condition of being the longest survivor.

But we have already seen that "the survival of the fittest," is, according to Mr. Darwin, a synonym for "natural selection"; hence it follows that "the conditions of existence" imply neither more nor less than what is implied by "natural selection" when this expression is properly explained, and may be used instead of it; so that when Mr. Darwin says that "natural selection" is the main but not exclusive means of modification, he must mean, consciously or unconsciously, that "the conditions of existence" are the main but not exclusive means of modification. But this is only falling in with "the views and erroneous grounds of opinion," as Mr. Darwin briefly calls them, of Lamarck himself; a fact which Mr. Darwin's readers would have seen more readily if he had kept to the use of the words "survival of the fittest" instead of "natural selection." Of that expression Mr. Darwin says[356] that it is "more accurate" than natural selection, but naively adds, "and sometimes equally convenient."

I have said that there is a practical identity of meaning between "natural selection" and "the conditions of existence," when both expressions are fully extended. I say this, however, without prejudice to my right of maintaining that, of the two expressions, the one is accurate, lucid, and calculated to keep the thread of the argument well in sight of the reader, while the other is inaccurate, and always, if I may say so, less "convenient," as being always liable to lead the reader astray. Nor should it be lost sight of that Lamarck and Dr. Erasmus Darwin maintain that species and genera have arisen *because animals can fashion themselves into accord with* their conditions, so that, as Lamarck is so continually insisting, the action of the conditions is indirect only--changed use and disuse being the direct causes; while, according to Mr. Darwin, it is natural selection itself (which, as we have seen, is but another way of saying conditions of existence) that is the most important means of modification.

The identity of meaning above insisted on was, on the face of it, almost as obscure as that between "*evêque* and bishop." Yet we know that "*evêque*" is "episc" and "bishop" "piscop," and that "episcopus" is the Latin for bishop; the words, therefore, are really one and the same, in spite of the difference in their appearance. I think I can show, moreover, that Mr. Darwin himself holds natural selection and the conditions of existence to be one and the same thing. For he writes, "in one sense," and it is hard to see any sense but one in what follows, "the conditions of life may be said not only to cause variability" (so that here Mr. Darwin appears to support Lamarck's main thesis) "either directly or indirectly, but likewise to include natural selection; for the conditions determine whether this or that variety shall survive."[357] But later on we find that "the expression of conditions of existence, so often insisted upon by the illustrious Cuvier" (and surely also by the illustrious Lamarck, though he calls them "*circonstances*") "is fully embraced by the principle of natural selection."[358] So we see that the conditions of life "*include*" natural selection, and yet the conditions of existence "*are fully embraced by*" natural selection, which, I take it, is an enigmatic way of saying that they are one and the same thing, for it is not until two bodies absolutely coincide and occupy the same space that the one can be said both to include and to be embraced by the other.

The difficulty, again, of understanding Mr. Darwin's meaning is enhanced by his repeatedly writing of "natural selection," or the fact that the fittest survive in the struggle for existence, as though it were the same thing as "evolution" or the descent, through the accumulation of small modifications in many successive generations, of one species from another and different one. In the concluding and recapitulatory chapter of the 'Origin of Species,' he writes:--

"Turning to geographical distribution, the difficulties encountered *on the theory of descent with modification* are serious enough;"[359] and in the next paragraph, "As, according to *the theory of natural selection, &c.*," the context showing that in each case descent with modification is intended.

Again:--

"On the theory of the *natural selection* of successive, slight, but profitable, modifications,"[360] that is to say, on the theory of the survival of the fittest; while on the next page we find "*the theory of descent with modification*," and "*the principle of natural selection*," used as though they were convertible terms.

Again:--

"The existence of closely allied or representative species in any two areas implies, *on the theory of descent with modification, &c.*;"[361] and, in the next paragraph, "*the theory of natural selection*, with its contingencies of extinction and divergence of character," is substituted as though the two expressions were identical.

This is calculated to mislead. Independently of the fact that "natural selection," or "the survival of the fittest," is in no sense a theory, but simply an observed fact, yet even if the words are allowed to stand for "descent with modification by means of natural selection," it is still misleading to write as though this were synonymous with "the theory of evolution," or "the theory of descent with modification." To do this prevents the reader from bearing in mind that "evolution by means of the circumstance-suiting power of plants and animals" as advanced by the earlier evolutionists; and "evolution by means of lucky accidents" with comparatively little circumstance-suiting power, are two very different things, of which the one may be true and the other untrue. It leads the reader to forget that evolution by no means stands or falls with evolution by means of natural selection, and makes him think that if he accepts evolution at all, he is bound to Mr. Darwin's view of it. Hence, when he falls in with such writers as Professor Mivart and the Rev. J. J. Murphy, who show, and very plainly, that the survival of the fittest, unsupplemented by something which shall give a definite aim to the variations which successively occur, fails to account for the coadaptations of need and structure, he imagines that evolution has much less to say for itself than it really has. If Mr. Darwin, instead of taking the line which he has thought fit to adopt towards Buffon, Dr. Erasmus Darwin, Lamarck, and the author of the 'Vestiges,' had shown us what these men taught, why they taught it, wherein they were wrong, and how he proposed to set them right, he would have taken a course at once more agreeable with ordinary practice, and more likely to clear misconception from his own mind and from those of his readers.

Mr. Darwin says,[362] "it is easy to hide our ignorance under such expressions as 'the plan of creation' and 'unity of design.'" Surely, also, it is easy to hide want of precision of thought, and the absence of any fundamental difference between his own main conclusion and that of Dr. Darwin and Lamarck whom he condemns, under the term "natural selection."

I assure the reader that I find the task of forming a clear, well-defined conception of Mr. Darwin's meaning, as expressed in his 'Origin of Species,' comparable only to that of one who has to act on the advice of a lawyer who has obscured the main issue as far as he can, and whose chief aim has been to make as many loopholes as possible for himself to escape through in case of his being called to account. Or, again, to that of one who has to construe an Act of Parliament which was originally framed so as to throw dust in the eyes of those who would oppose the measure, and which, having been since found unworkable, has had clauses repealed

and inserted up and down it, till it is in an inextricable tangle of confusion and contradiction.

As an example of my meaning, I will quote a passage to which I called attention in 'Life and Habit.' It runs:--

"In the earlier editions of this work I underrated, as now seems probable, the frequency and importance of modifications due to spontaneous variability. But it is impossible to attribute to *this cause*" (i. e. spontaneous variability, which is itself only an expression for unknown causes) "the innumerable structures which are so well adapted to the habits of life of each species. I can no more believe in *this*" (i. e. that the innumerable structures, &c., can be due to unknown causes) "than that the well adapted form of a racehorse or greyhound, which, before the principle of selection by man was well understood, excited so much surprise in the minds of the older naturalists, can *thus*" (i. e. by attributing them to unknown causes) "be explained."[363]

This amounts to saying that unknown causes can do so much, but cannot do so much more. On this passage I wrote, in 'Life and Habit':--

"It is impossible to believe that, after years of reflection upon his subject, Mr. Darwin should have written as above, especially in such a place, if his mind was clear about his own position. Immediately after the admission of a certain amount of miscalculation there comes a more or less exculpatory sentence, which sounds so right that ninety-nine people out of a hundred would walk through it, unless led by some exigency of their own position to examine it closely, but which yet, upon examination, proves to be as nearly meaningless as a sentence can be."[364]

No one, to my knowledge, has impugned the justice of this criticism, and I may say that further study of Mr. Darwin's works has only strengthened my conviction of the confusion and inaccuracy of thought, which detracts so greatly from their value.

So little is it generally understood that "evolution" and what is called "Darwinism" convey indeed the same main conclusion, but that this conclusion has been reached by two distinct roads, one of which is impregnable, while the other has already fallen into the hands of the enemy, that in the last November number of the 'Nineteenth Century' Professor Tyndall, while referring to descent with modification or evolution, speaks of it as though it were one and inseparable from Mr. Darwin's theory that it has come about mainly by means of natural selection. He writes:--

"*Darwin's theory*, as pointed out nine or ten years ago by Helmholtz and Hooker, was then exactly in this condition of growth; and had they to speak of the subject to-day they would be able to announce an enormous strengthening of the theoretic fibre. Fissures in continuity which then existed, and which left little hope of being ever spanned, have been since bridged over, so that the further *the theory* is tested the more fully does it harmonize with progressive experience and discovery. We shall never probably fill all the gaps; but this will not prevent a profound belief in the truth of *the theory* from taking root in the general mind. Much less will it justify a total denial of *the theory*. The man of science, who assumes in such a case the position of a denier, is sure to be stranded and isolated."

This is in the true vein of the professional and orthodox scientist; of that new orthodoxy which is clamouring for endowment, and which would step into the Pope's shoes to-morrow, if we would only let it. If Professor Tyndall means that those who deny evolution will find themselves presently in a very small minority, I agree with him; but if he means that evolution is Mr. Darwin's theory, and that he who rejects what Mr. Darwin calls "the theory of natural selection" will find himself stranded, his assertion will pass muster with those only who know little of the history and literature of evolution.

FOOTNOTES:

[347] 'Origin of Species,' Hist. Sketch, p. xiii.

[348] 'Physical Basis of Mind,' p. 108.

[349] 'Origin of Species,' p. 146.

[350] Ibid. p. 75.

[351] Ibid. p. 88.

[352] 'Origin of Species,' p. 98.

[353] Ibid. p. 66.

[354] 'Physical Basis of the Mind,' p. 109, 1878.

[355] 'Physical Basis of the Mind,' p. 107, 1878.

[356] 'Origin of Species,' p. 49.

[357] 'Origin of Species,' p. 107.

[358] Ibid. p. 166.

[359] 'Origin of Species,' p. 406.

[360] Ibid, p. 416.

[361] Ibid. p. 419.

[362] 'Origin of Species,' p. 422.

[363] 'Origin of Species,' p. 171, ed. 1876.

[364] 'Life and Habit,' p. 260.

CHAPTER XXI.

MR. DARWIN'S DEFENCE OF THE EXPRESSION, NATURAL SELECTION--PROFESSOR MIVART AND NATURAL SELECTION.

So important is it that we should come to a clear understanding upon the positions taken by Mr. Darwin and Lamarck respectively, that at the risk of wearying the reader I will endeavour to exhaust this subject here. In order to do so, I will follow Mr. Darwin's answer to those who have objected to the expression, "natural selection."

Mr. Darwin says:--

"Several writers have misapprehended or objected to the term 'natural selection.' Some have even imagined that natural selection induces variability."[365]

And small wonder if they have; but those who have fallen into this error are hardly worth considering. The true complaint is that Mr. Darwin has too often written of "natural selection" as though it does induce variability, and that his language concerning it is so confusing that the reader is not helped to see that it really comes to nothing but a cloak of difference from his predecessors, under which there lurks a concealed identity of opinion as to the main facts. The reader is thus led to look upon it as something positive and special, and, in spite of Mr. Darwin's disclaimer, to think of it as an actively efficient cause.

Few will deny that this complaint is a just one, or that ninety-nine out of a hundred readers of average intelligence, if asked, after reading Mr. Darwin's 'Origin of Species,' what was the most important cause of modification, would answer "natural selection." Let the same readers have read the 'Zoonomia' of Dr. Erasmus Darwin, or the 'Philosophie Zoologique' of Lamarck, and they would at once reply, "the wishes of an animal or plant, as varying with its varying conditions," or more briefly, "sense of need."

"Whereas," continues Mr. Darwin, "it" (natural selection) "implies only the preservation of such variations as arise, and are beneficial to the being under its conditions of life. No one objects to agriculturists speaking of the potent effects of man's selection."

Of course not; for there *is* an actual creature man, who actually does select with a set purpose in order to produce such and such a result, which result he presently produces.

"And in this case the individual differences given by nature, which man for some object selects, must first occur."

This shows that the complaint has already reached Mr. Darwin, that in not showing us how "the individual differences first occur," he is really leaving us absolutely in the dark as to the cause of all modification--giving us an 'Origin of Species' with "the origin" cut out; but I do not think that any reader who has not been compelled to go somewhat deeply into the question would find out that this is the real gist of the objection which Mr. Darwin is appearing to combat. A general impression is left upon the reader that some very foolish objectors are being put to silence, that Mr. Darwin is the most candid literary opponent in the world, and as just as Aristides himself; but if the unassisted reader will cross-question himself what it is all about, I shall be much surprised if he is ready with his answer.

"Others"--to resume our criticism on Mr. Darwin's defence--"have objected that the term implies conscious choice in the animals which become modified, and it has been even urged that as plants have no volition, natural selection is not applicable to them!"

This--unfortunately--must have been the objection of a slovenly, or wilfully misapprehending reader, and was unworthy of serious notice. But its introduction here tends to draw the reader from the true ground of complaint, which is that at the end of Mr. Darwin's book we stand much in the same place as we did when we started, as regards any knowledge of what is the "origin of species."

"In the literal sense of the word, no doubt, natural selection is a false term."

Then why use it when another, and, by Mr. Darwin's own admission, a "more accurate" one is to hand in "the survival of the fittest"?[366] This term is not appreciably longer than natural selection. Mr. Darwin may say, indeed, that it is "sometimes" as convenient a term as natural selection; but the kind of men who exercise permanent effect upon the opinions of other people will bid such a passage as this stand aside somewhat sternly. If a term is not appreciably longer than another, and if at the same time it more accurately expresses the idea which is intended to be conveyed, it is not sometimes only, but always, more convenient, and should immediately be substituted for the less accurate one.

No one complains of the use of what is, strictly speaking, an inaccurate expression, when it is nevertheless the best that we can get. It may be doubted whether there is any such thing possible as a perfectly accurate expression. All words that are not simply names of things are apt to turn out little else than compendious false analogies; but we have a right to complain when a writer tells us that he is using a less accurate expression when a more accurate one is ready to his hand. Hence, when Mr. Darwin continues, "Who ever objected to chemists speaking of the elective affinities of the various elements? and yet an acid cannot strictly be said to elect the base with which it by preference combines," he is beside the mark. Chemists do not speak of "elective affinities" in spite of there being a more accurate and not appreciably longer expression at their disposal.

"It has been said," continues Mr. Darwin, "that I speak of natural selection as an active power or deity. But who objects to an author speaking of the attraction of gravity? Everyone knows what is meant and implied by such metaphorical expressions, and they are almost necessary for brevity."

Mr. Darwin certainly does speak of natural selection "acting," "accumulating," "operating"; and if "every-one knew what was meant and implied by this metaphorical expression," as they now do, or think they do, in the case of the attraction of gravity, there might be less ground of complaint; but the expression was known to very few at the time Mr. Darwin introduced it, and was used with so much ambiguity, and with so little to protect the reader from falling into the error of supposing that it was the cause of the modifications which we see around us, that we had a just right to complain, even in the first instance; much more should we do so on the score of the retention of the expression when a more accurate one had been found.

If the "survival of the fittest" had been used, to the total excision of "natural selection" from every page in Mr. Darwin's book--it would have been easily seen that "the survival of the fittest" is no more a cause of modification, and hence can give no more explanation concerning the origin of species, than the fact of a number of competitors in a race failing to run the whole course, or to run it as quickly as the winner, can explain how the winner came to have good legs and lungs. According to Lamarck, the winner will have got these by means of sense of need, and consequent practice and training, on his own part, and on that of his forefathers; according to Mr. Darwin, the "most important means" of his getting them is his "happening" to be born with them, coupled, with the fact that his uncles and aunts for many generations could not run so well as his ancestors in the direct line. But can the fact of his uncles and aunts running less well than his fathers and mothers be a means of his fathers and mothers coming to run *better than they used to run*?

If the reader will bear in mind the idea of the runners in a race, it will help him to see the point at issue between Mr. Darwin and Lamarck. Perhaps also the double meaning of the word race, as expressing equally a breed and a competition, may not be wholly without significance. What we want to be told is, not that a runner will win the prize if he can run "ever such a little" faster than his fellows--we know this--but by what process he comes to be able to run ever such a little faster.

"So, again," continues Mr. Darwin, "it is difficult to avoid personifying nature, but I mean by nature only the aggregate action and product of many natural laws, and by laws the sequence of events as ascertained by us."

This, again, is raising up a dead man in order to knock him down. Nature has been personified for more than two thousand years, and every one understands that nature is no more really a woman than hope or justice, or than God is like the pictures of the mediæval painters; no one whose objection was worth notice could have objected to the personification of nature.

Mr. Darwin concludes:--

"With a little familiarity, such superficial objections will be forgotten."[367]

As a matter of fact, I do not see any greater tendency to acquiesce in Mr. Darwin's claim on behalf of natural selection than there was a few years ago, but on the contrary, that discontent is daily growing. To say nothing of the Rev. J. J. Murphy and Professor Mivart, the late Mr. G. H. Lewes did not find the objection a superficial one, nor yet did he find it disappear "with a little familiarity"; on the contrary, the more familiar he became with it the less he appeared to like it. I may even go, without fear, so far as to say that any writer who now uses the expression "natural selection," writes himself down thereby as behind the age. It is with great pleasure that I observe Mr. Francis Darwin in his recent lecture[368] to have kept clear of it altogether, and to have made use of no expression, and advocated no doctrine to which either Dr. Erasmus Darwin or Lamarck would not have readily assented. I think I may affirm confidently that a few years ago any such lecture would have contained repeated reference to Natural Selection. For my own part I know of few passages in any theological writer which please me less than the one which I have above followed sentence by sentence. I know of few which should better serve to show us the sort of danger we should run if we were to let men of science get the upper hand of us.

Natural Selection, then, is only another way of saying "Nature." Mr. Darwin seems to be aware of this when he writes, "Nature, if I may be allowed to personify the natural preservation or survival of the fittest." And again, at the bottom of the same page, "It may metaphorically be said that *natural selection is daily and hourly scrutinizing* throughout the world the slightest variations."[369] It may be metaphorically said that *Nature* is daily and hourly scrutinizing, but it cannot be said consistently with any right use of words, metaphorical or otherwise, that natural selection scrutinizes, unless natural selection is merely a somewhat cumbrous synonym for Nature. When, therefore, Mr. Darwin says that natural selection is the "most important, but not the exclusive means" whereby any modification has been effected, he is really saying that Nature is the most important means of modification--which is only another way of telling us that variation causes variations, and is all very true as far as it goes.

I did not read Professor Mivart's 'Lessons from Nature,' until I had written all my own criticism on Mr. Darwin's position. From that work, however, I now quote the following:--

"It cannot then be contested that the far-famed 'Origin of Species,' that, namely, by 'Natural Selection,' has been repudiated in fact, though not expressly even by its own author. This circumstance, which is simply undeniable, might dispense us from any further consideration of the hypothesis itself. But the "conspiracy of silence," which has accompanied the repudiation tends to lead the unthinking many to suppose that the same importance still attaches to it as at first. On this account it may be well to ask the question, what, after all, *is* 'Natural Selection'?

"The answer may seem surprising to some, but it is none the less true, that 'Natural Selection' is simply nothing. It is an apparently positive name for a really negative effect, and is therefore an eminently misleading term. By 'Natural Selection' is meant the result of all the destructive agencies of Nature, destructive to individuals and to races by destroying their lives or their powers of propagation. Evidently, *the cause of the distinction of species* (supposing such distinction to be brought about in natural generation) *must be that which causes variation, and variation in one determinate direction in at least several individuals simultaneously*." I should like to have added here the words "and during many successive generations," but they will go very sufficiently without saying.

"At the same time," continues Professor Mivart, "it is freely conceded that the destructive agencies in nature do succeed in preventing the perpetuation of monstrous, abortive, and feeble attempts at the performance of the evolutionary process, that they rapidly remove antecedent forms when new ones are evolved more in harmony with surrounding conditions, and that their action results in the formation of new characters when these have once attained sufficient completeness to be of real utility to their possessor.

"Continued reflection, and five years further pondering over the problems of specific origin have more and more convinced me that the conception, that the origin of all species 'man included' is due simply to conditions which are (to use Mr. Darwin's own words) 'strictly accidental,' is a conception utterly irrational."

"With regard to the conception as now put forward by Mr. Darwin, I cannot truly characterize it but by an epithet which I employ only with much reluctance. I weigh my words and have present to my mind the many distinguished naturalists who have accepted the notion, and yet I cannot hesitate to call it a '*puerile hypothesis*.'"[370]

I am afraid I cannot go with Professor Mivart farther than this point, though I have a strong feeling as though his conclusion is true, that "the material universe is always and everywhere sustained and directed by an infinite cause, for which to us the word mind is the least inadequate and misleading symbol." But I feel that any attempt to deal with such a question is going far beyond that sphere in which man's powers may be at present employed with advantage: I trust, therefore, that I may never try to verify it, and am indifferent whether it is correct or not.

Again, I should probably differ from Professor Mivart in finding this mind inseparable from the material universe in which we live and move. So that I could neither conceive of such a mind influencing and directing the universe from a point as it were outside the universe itself, nor yet of a universe as existing without there being present--or having been present--in its every particle something for which mind should be the least inadequate and misleading symbol. But the subject is far beyond me.

As regards Professor Mivart's denunciations of natural selection, I have only one fault to find with them, namely, that they do not speak out with sufficient bluntness. The difficulty of showing the fallacy of Mr. Darwin's position, is the difficulty of grasping a will-o'-the-wisp. A concluding example will put this clearly before the reader, and at the same time serve to illustrate the most tangible feature of difference between Mr. Darwin and Lamarck.

FOOTNOTES:

[365] 'Origin of Species,' p. 62.

[366] 'Origin of Species,' p. 49.

[367] 'Origin of Species,' p. 63.

[368] 'Nature,' March 14 and 21, 1878.

[369] 'Origin of Species,' p. 65.

[370] 'Lessons from Nature,' p. 300.

CHAPTER XXII.

THE CASE OF THE MADEIRA BEETLES AS ILLUSTRATING THE DIFFERENCE BETWEEN THE EVOLUTION OF LAMARCK AND OF MR. CHARLES DARWIN--CONCLUSION.

An island of no very great extent is surrounded by a sea which cuts it off for many miles from the nearest land. It lies a good deal exposed to winds, so that the beetles which live upon it are in continual danger of being blown out to sea if they fly during the hours and seasons when the wind is blowing. It is found that an unusually large proportion of the beetles inhabiting this island are either without wings or have their wings in a useless and merely rudimentary state; and that a large number of kinds which are very common on the nearest mainland, but which are compelled to use their wings in seeking their food, are here entirely wanting. It is also observed that the beetles on this island generally lie much concealed until the wind lulls and the sun shines. These are the facts; let us now see how Lamarck would treat them.

Lamarck would say that the beetles once being on this island it became one of the conditions of their existence that they should not get blown out to sea. For once blown out to sea, they would be quite certain to be drowned. Beetles, when they fly, generally fly for some purpose, and do not like having that purpose interfered with by something which can carry them all-whithers, whether they like it or no. If they are flying and find the wind taking them in a wrong direction, or seaward--

which they know will be fatal to them--they stop flying as soon as may be, and alight on *terra firma*. But if the wind is very prevalent the beetles can find but little opportunity for flying at all: they will therefore lie quiet all day and do as best they can to get their living on foot instead of on the wing. There will thus be a long-continued disuse of wings, and this will gradually diminish the development of the wings themselves, till after a sufficient number of generations these will either disappear altogether, or be seen in a rudimentary condition only. For each beetle which has made but little use of its wings will be liable to leave offspring with a slightly diminished wing, some other organ which has been used instead of the wing becoming proportionately developed. It is thus seen that the conditions of existence are the indirect cause of the wings becoming rudimentary, inasmuch as they preclude the beetles from using them; the disuse however on the part of the beetles themselves is the direct cause.

Now let us see how Mr. Darwin deals with the same case. He writes:--

"In some cases we might easily set down to disuse, modifications of structure which are *wholly* or *mainly* due to natural selection." Then follow the facts about the beetles of Madeira, as I have given them above. While we are reading them we naturally make up our minds that the winglessness of the beetles will prove due either wholly, or at any rate mainly, to natural selection, and that though it would be easy to set it down to disuse, yet we must on no account do so. The facts having been stated,

Mr. Darwin continues:--"These several considerations make me believe that the wingless condition of so many Madeira beetles is mainly due to the action of natural selection," and when we go on to the words that immediately follow, "combined probably with disuse," we are almost surprised at finding that disuse has had anything to do with the matter. We feel a languid wish to know exactly how much and in what way it has entered into the combination; but we find it difficult to think the matter out, and are glad to take it for granted that the part played by disuse must be so unimportant that we need not consider it. Mr. Darwin continues:--

"For during many successive generations each individual beetle which flew least, either from its wings having been ever so little less perfectly developed, or from indolent habit, will have had the best chance of surviving from not having been blown out to sea; and on the other hand those beetles which most readily took to flight would oftenest be blown out to sea and perish."[371]

So apt are we to believe what we are told, when it is told us gravely and with authority, and when there is no statement at hand to contradict it, that we fail to see that Mr. Darwin is all the time really attributing the winglessness of the Madeira beetles either to the *quâ* him *unknown causes* which have led to the "ever so little less perfect development of wing" on the part of the beetles that leave offspring--that is to say, is admitting that he can give no account of the matter--or else to the "indolent habit" of the parent beetles which has led them to disuse their wings, and hence gradually to lose them--which is neither more nor less than the "erroneous grounds of opinion," and "well-known doctrine" of Lamarck.

For Mr. Darwin cannot mean that the fact of some beetles being blown out to sea is the most important means whereby certain other beetles come to have smaller wings--that the Madeira beetles in fact come to have smaller wings mainly because their large winged uncles and aunts--go away.

But if he does not mean this, what becomes of natural selection?

For in this case we are left exactly where Lamarck left us, and must hold that such beetles as have smaller wings have them because the conditions of life or "circumstances" in which their parents were placed, rendered it inconvenient to them to fly, and thus led them to leave off using their wings.

Granted, that if there had been nothing to take unmodified beetles away, there would have been less room and scope for the modified beetles; also that unmodified beetles would have intermixed with the modified, and impeded the prevalence of the modification. But anything else than such removal of unmodified individuals would be contrary to our hypothesis. The very essence of conditions of existence is that there *shall be* something to take away those which do not comply with the conditions; if there is nothing to render such and such a course a *sine quâ non* for life, there is no condition of existence in respect of this course, and no modification according to Lamarck could follow, as there would be no changed distribution of use.

I think that if I were to leave this matter here I should have said enough to make the reader feel that Lamarck's system is direct, intelligible and sufficient--while Mr. Darwin's is confused and confusing. I may however quote Mr. Darwin himself as throwing his theory about the Madeira beetles on one side in a later passage, for he writes:--

"It is probable that *disuse has been the main agent in rendering organs rudimentary*," or in other words that Lamarck was quite right--nor does one see why if disuse is after all the main agent in rendering an organ rudimentary, use should not have been the main agent in developing it--but let that pass. "It (disuse) would at first lead," continues Mr. Darwin, "by slow steps to the more and more complete reduction of a part, until at last it became rudimentary--as in the case of the eyes of animals inhabiting dark caverns, and of the wings of birds inhabiting oceanic islands, which have seldom been forced by beasts of prey to take flight, and have ultimately lost the power of flying. Again, an organ useful under certain conditions, might become injurious under others, *as with the wings of beetles living on small and exposed islands*;"[372] so that the rudimentary condition of the Madeira beetles' wings is here set down as mainly due to disuse--while above we find it mainly due to natural selection--I should say that immediately after the word "islands" just quoted, Mr. Darwin adds "and in this case natural selection will have aided in reducing the organ, until it was rendered harmless and rudimentary," but this is Mr. Darwin's manner, and must go for what it is worth.

How refreshing to turn to the simple straightforward language of Lamarck.

"Long continued disuse," he writes, "in consequence of the habits which an animal has contracted, gradually reduces an organ, and leads to its final disappearance....

"Eyes placed in the head form an essential part of that plan on which we observe all vertebrate organisms to be constructed. Nevertheless the mole which uses its vision very little, has eyes which are only very small and hardly apparent.

"The *aspalax* of Olivier, which lives underground like the mole, and exposes itself even less than the mole to the light of day, has wholly lost the use of its sight, nor does it retain more than mere traces of visual organs, these traces again being hidden under the skin and under certain other parts which cover them up and leave not even the smallest access to the light. The Proteus, an aquatic reptile akin to the Salamander and living in deep and obscure cavities under water, has, like the aspalax, no longer anything but traces of eyes remaining--traces which are again entirely hidden and covered up.[373]

"The following consideration should be decisive.

"Light cannot penetrate everywhere, and as a consequence, animals which live habitually in places which it cannot reach, do not have an opportunity of using eyes, even though they have got them; but animals which form part of a system of organization which comprises eyes as an invariable rule among its organs, must have had eyes originally. Since then we find among these animals some which have lost their eyes, and which have only concealed traces of these organs, it is evident that the impoverishment, and even disappearance of the organs in question, must be the effect of long-continued disuse.

"A proof of this is to be found in the fact that the organ of hearing is never in like case with that of sight; we always find it in animals of whose system of organization hearing is a component part; and for the following reason, namely, that sound, which is the effect of vibration upon the ear, can penetrate everywhere, and pass even through massive intermediate bodies. Any animal, therefore, with an organic system of which the ear is an essential part, can always find a use for its ears, no matter where it inhabits. We never, therefore, come upon rudimentary ears among the vertebrata, and when, going down the scale of life lower than the vertebrata, we come to a point at which the ear is no longer to be found; we never come upon ears again in any lower class.

"Not so with the organ of sight: we see this organ disappear, reappear, and disappear again with the possibility or impossibility of using eyes on the part of the creature itself.[374]

"The great development of mantle in the acephalous

molluscs has rendered eyes, and even a head, entirely useless to them. These organs, though belonging to the type of the organism, and by rights included in it, have had to disappear and become annihilated owing to continued default of use.

"Many insects which, by the analogy of their order and even genus, should have wings, have nevertheless lost them more or less completely through disuse. A number of coleoptera, orthoptera, hymenoptera, and hemiptera give us examples, the habits of these animals never leading them to use their wings."[375]

I will here bring this present volume to a conclusion, hoping, however, to return to the same subject shortly, but to that part of it which bears upon longevity and the phenomena of old age. In 'Life and Habit' I pointed out that if differentiations of structure and instinct are considered as due to the different desires under different circumstances of an organism, which must be regarded as a single creature, though its development has extended over millions of years, and which is guided mainly by habit and memory until some disturbing cause compels invention--then the longevity of each generation or stage of this organism should depend upon the lateness of the average age of reproduction in each generation; so that an organism (using the word in its usual signification) which did not upon the average begin to reproduce itself till it was twenty, should be longer lived than one that on the average begins to reproduce itself at a year old. I also maintained that the phenomena of old age should be referred to failure of memory on the part of the organism, which in the embryonic stages, infancy, youth, and early manhood, leans upon the memory of what it did when it was in the persons of its ancestors; in middle life, carries its action onward by means of the impetus, already received, and by the force of habit; and in old age becomes puzzled, having no experience of any past existence at seventy-five, we will say, to guide it, and therefore forgetting itself more and more completely till it dies. I hope to extend this, and to bring forward arguments in support of it in a future work.

Of the importance of the theory put forward in 'Life and Habit'--I am daily more and more convinced. Unless we admit oneness of personality between parents and offspring, memory of the often repeated facts of past existences, the latency of that memory until it is rekindled by the presence of the associated ideas, or of a sufficient number of them, and the far-reaching consequences of the unconsciousness which results from habitual action, evolution does not greatly add to our knowledge as to how we shall live here to the best advantage. Add these considerations, and its value as a guide becomes immediately apparent; a new light is poured upon a hundred problems of the greatest delicacy and difficulty. Not the least interesting of these is the gradual extension of human longevity--an extension, however, which cannot be effected till many many generations as yet unborn have come and gone. There is nothing, however, to prevent man's becoming as long lived as the oak if he will persevere for many generations in the steps which can alone lead to this result. Another interesting achievement which should be more quickly attainable, though still not in our own time, is the earlier maturity of those animals whose rapid maturity is an advantage to us, but whose longevity is not to our purpose.

The question--Evolution or Direct Creation of all species?--has been settled in favour of Evolution. A hardly less interesting and important battle has now to be fought over the question whether we are to accept the evolution of the founders of the theory--with the adjuncts hinted at by Dr. Darwin and Mr. Matthew, and insisted on, so far as I can gather, by Professor Hering and myself--or the evolution of Mr. Darwin, which denies the purposiveness or teleology inherent in evolution as first propounded. I am assured that such of my readers as I can persuade to prefer the old evolution to the new will have but little reason to regret their preference.

P.S.--As these sheets leave my hands, my attention is called to a review of Professor Haeckel's 'Evolution of Man,' by Mr. A. E. Wallace, in the 'Academy' for April 12, 1879. "Professor Haeckel maintains," says Mr. Wallace, "*that the struggle for existence in nature evolves new forms without design, just as the will of man produces new varieties in cultivation with design*." I maintain in preference with the older evolutionists, that in consequence of change in the conditions of their existence, *organisms design new forms for themselves, and carry those designs out in additions to, and modifications of, their own bodies*.

"The science of rudimentary organs," continues Mr.

Wallace, "which Haeckel terms 'dysteleology, or the doctrine of purposelessness,' is here discussed, and a number of interesting examples are given, the conclusion being that they prove the mechanical or monistic conception of the origin of organisms to be correct, and the idea of any 'all-wise creative plan an ancient fable.'" I see no reason to suppose, or again not to suppose, an all-wise creative plan. I decline to go into this question, believing it to be not yet ripe, nor nearly ripe, for consideration. I see purpose, however, in rudimentary organs as much as in useful ones, but a spent or extinct purpose--a purpose which has been fulfilled, and is now forgotten--the rudimentary organ being repeated from force of habit, indolence, and dislike of change, so long as it does not, to use the words of Buffon, "stand in the way of the fair development" of other parts which are found useful and necessary. I demur, therefore, to the inference of "purposelessness" which I gather that Professor Haeckel draws from these organs.

In the 'Academy' for April 19, 1879, Mr. Wallace quotes Professor Haeckel as saying that our "highly purposive and admirably-constituted sense-organs have developed without premeditated aim; that they have originated by the same mechanical process of Natural Selection, by the same constant interaction of Adaptation and Heredity [what *is* Heredity but another word for unknown causes, unless it is explained in some such manner as in 'Life and Habit'?] by which all the other purposive contrivances of the animal organization have been slowly and gradually evolved during the struggle for existence."

I see no evidence for "premeditated aim" at any modification very far in advance of an existing organ, any more than I do for "premeditated aim" on man's part at any as yet inconceivable mechanical invention; but as in the case of man's inventions, so also in that of the organs of animals and plants, modification is due to the accumulation of small, well-considered improvements, as found necessary in practice, and the conduct of their affairs. Each step having been purposive, the whole road has been travelled purposively; nor is the purposiveness of such an organ, we will say, as the eye, barred by the fact that invention has doubtless been aided by some of those happy accidents which from time to time happen to all who keep their wits about them, and know how to turn the gifts of Fortune to account.

FOOTNOTES:

[371] 'Origin of Species,' p. 109.

[372] 'Origin of Species, p. 401.

[373] 'Phil. Zool.,' tom. i. p. 242.

[374] 'Phil. Zool.,' tom. i. p. 244.

[375] 'Phil. Zool.,' tom. i. p. 245.

APPENDIX.

CHAPTER I.

REVIEWS OF 'EVOLUTION, OLD AND NEW.'

Those who have been at the pains to read the foregoing book will, perhaps, pardon me if I put before them a short account of the reception it has met with: I will not waste time by arguing with my critics at any length; it will be enough if I place some of their remarks upon my book under the same cover as the book itself, with here and there a word or two of comment.

The only reviews which have come under my notice appeared in the 'Academy' and the 'Examiner,' both of May 17, 1879; the 'Edinburgh Daily Review,' May 23, 1879; 'City Press,' May 21, 1879; 'Field,' May 26, 1879; 'Saturday Review,' May 31, 1879; 'Daily Chronicle,' May 31, 1879; 'Graphic' and 'Nature,' both June 12, 1879; 'Pall Mall Gazette,' June 18, 1879; 'Literary World,' June 20, 1879; 'Scotsman,' June 24, 1879; 'British Journal of Homoeopathy' and 'Mind,' both July 1, 1879; 'Journal of Science,' July 18, 1879; 'Westminster Review,' July, 1879; 'Athenæum,' July 26, 1879; 'Daily News,' July 29, 1879; 'Manchester City News,' August 16, 1879; 'Nonconformist,' November 26, 1879; 'Popular Science Review,' Jan. 1, 1880; 'Morning Post,' Jan. 12, 1880.

Some of the most hostile passages in the reviews above referred to are as follows:--

"From beginning to end, our eccentric author treats us to a dazzling flood of epigram, invective, and what appears to be argument; and finally leaves us without a single clear idea as to what he has been driving at."

"Mr. Butler comes forward, as it were, to proclaim himself a professional satirist, and a mystifier who will do his best to leave you utterly in the dark with regard to his system of juggling. Is he a teleological theologian making fun of evolution? Is he an evolutionist making fun of teleology? Is he a man of letters making fun of science? Or is he a master of pure irony making fun of all three, and of his audience as well? For our part we decline to commit ourselves, and prefer to observe, as Mr. Butler observes of Von Hartmann, that if his meaning is anything like what he says it is, we can only say that it has not been given us to form any definite conception whatever as to what that meaning may be."--'Academy,' May 17, 1879, Signed Grant Allen.

Here is another criticism of "Evolution, Old and New"--also, I believe I am warranted in saying, by Mr. Grant Allen. These two criticisms appeared on the same day; how many more Mr. Allen may have written later on I do not know.

We find the writer who in the 'Academy' declares that he has been left without "a single clear idea" as to what 'Evolution, Old and New,' has been driving at saying on the same day in the 'Examiner' that 'Evolution, Old and New,' "has a more evident purpose than any of its predecessors." If so, I am afraid the predecessors must have puzzled Mr. Allen very unpleasantly. What the purpose of 'Evolution, Old and New,' is, he proceeds to explain:--

"As to his (Mr. Butler's) main argument, it comes briefly to this: natural selection does not originate favourable varieties, it only passively permits them to exist; therefore it is the unknown cause which produced the variations, not the natural selection which spared them, that ought to count as the mainspring of evolution. That unknown cause Mr. Butler boldly declares to be the will of the organism itself. An intelligent ascidian wanted a pair of eyes,[376] so set to work and made itself a pair, exactly as a man makes a microscope; a talented fish conceived the idea of walking on dry land, so it developed legs, turned its swim bladder into a pair of lungs, and became an amphibian; an æsthetic guinea-fowl admired bright colours, so it bought a paint-box, studied Mr. Whistler's ornamental designs, and, painting itself a gilded and ocellated tail, was thenceforth a peacock. But how about plants? Mr. Butler does not shirk even this difficulty. The theory must be maintained at all hazards.... This is the sort of mystical nonsense from which we had hoped Mr. Darwin had for ever saved us."--'Examiner,' May 17, 1879.

In this last article, Mr. Allen has said that I am a man of genius, "with the unmistakable signet-mark upon my forehead." I have been subjected to a good deal of obloquy and misrepresentation at one time or another, but this passage by Mr. Allen is the only one I have seen that has made me seriously uneasy about the prospects of my literary reputation.

I see Mr. Allen has been lately writing an article in the 'Fortnightly Review' on the decay of criticism. Looking over it somewhat hurriedly, my eye was arrested by the following:--

"Nowadays any man can write, because there are papers enough to give employment to everybody. No reflection, no deliberation, no care; all is haste, fatal facility, stock phrases, commonplace ideas, and a ready pen that can turn itself to any task with equal ease, because supremely ignorant of all alike."

"The writer takes to his craft nowadays, not because he has taste for literature, but because he has an incurable faculty for scribbling. He has no culture, and he soon loses the power of taking pains, if he ever possessed it. But he can talk with glib superficiality and imposing confidence about every conceivable subject, from a play or a picture to a sermon or a metaphysical essay. It is the utter indifference to subject-matter, joined with the vulgar unscrupulousness of pretentious ignorance, that strikes the keynote of our existing criticism. Men write without taking the trouble to read or think."[377]

The 'Saturday Review' attacked 'Evolution, Old and New,' I may almost say savagely. It wrote: "When Mr. Butler's 'Life and Habit' came before us, we doubted whether his ambiguously expressed speculations belonged to the regions of playful but possibly scientific imagination, or of unscientific fancies; and we gave him the benefit of the doubt. In fact, we strained a point or two to find a reasonable meaning for him. He has now settled the question against himself. Not professing to have any particular competence in biology, natural history, or the scientific study of evidence in any shape whatever, and, indeed, rather glorying in his freedom from any such superfluities, he undertakes to assure the overwhelming majority of men of science, and the educated public who have followed their lead, that, while they have done well to be converted to the doctrine of the evolution and transmutation of species, they have been converted on entirely wrong grounds."

"When a writer who has not given as many weeks to the subject as Mr. Darwin has given years [as a matter of fact, it is now twenty years since I began to publish on the subject of Evolution] is not content to air his own crude, though clever, fallacies, but presumes to criticize Mr. Darwin with the superciliousness of a young schoolmaster looking over a boy's theme, it is difficult not to take him more seriously than he deserves or perhaps desires. One would think that Mr. Butler was the travelled and laborious observer of Nature, and Mr. Darwin the pert speculator, who takes all his facts at secondhand."

"Let us once more consider how matters stood a year or two before the 'Origin of Species' first appeared. The continuous evolution of animated Nature had in its favour the difficulty of drawing fixed lines between species and even larger divisions, all the indications of comparative anatomy and embryology, and a good deal of general scientific presumption. Several well-known writers, and some eminent enough to command respect, had expressed their belief in it. One or two far-seeing thinkers, among whom the place of honour must be assigned to Mr. Herbert Spencer, had done more. They had used their philosophic insight, which, to science, is the eye of faith, to descry the promised land almost within reach; they knew and announced how rich and spacious the heritage would be, if once the entry could be made good. But on that 'if' everything hung. Nature was not bound to give up her secret, or was bound only in a mocking covenant with an impossible condition: *Si cælum digito tetigeris*; if only some fortunate hand could touch the inaccessible firmament, and bring down the golden chain to earth! But fruition seemed out of sight. Even those who were most willing to advance in this direction, could only regret that they saw no road clear. There was a tempting vision, but nothing proven--many would have said nothing provable. A few years passed, and all this was changed. The doubtful speculation had become a firm and connected theory. In the room of scattered foragers and scouts, there was an irresistibly advancing column. Nature had surrendered her stronghold, and was disarmed of her secret. And if we ask who were the men by whom this was done, the answer is notorious, and there is but one answer possible: the names that are for ever associated with this great triumph are those of Charles Darwin and Wallace."[378]

I gave the lady or gentleman who wrote this an opportunity of acknowledging the authorship; but she or he preferred, not I think unnaturally, to remain anonymous.

The only other criticism of 'Evolution, Old and New,' to which I would call attention, appeared in 'Nature,' in a review of 'Unconscious Memory,' by Mr. Romanes, and contained the following passages:--

"But to be serious, if in charity we could deem Mr. Butler a lunatic, we should not be unprepared for any aberration of common sense that he might display.... A certain nobody writes a book ['Evolution, Old and New'] accusing the most illustrious man in his generation of burying the claims of certain illustrious predecessors out of the sight of all men. In the hope of gaining some notoriety by deserving, and perhaps receiving a contemptuous refutation from the eminent man in question, he publishes this book which, if it deserved serious consideration, would be not more of an insult to the particular man of science whom it accuses of conscious and wholesale plagiarism [there is no such accusation in 'Evolution, Old and New'] than it would be to men of science in general for requiring such elementary instruction on some of the most famous literature in science from an upstart ignoramus, who, until two or three years ago, considered himself a painter by profession."--'Nature,' Jan. 27, 1881.

In a subsequent letter to 'Nature,' Mr. Romanes said he had been "acting the part of policeman" by writing as he had done. Any unscrupulous reviewer may call himself a policeman if he likes, but he must not expect those whom he assails to recognize his pretensions. 'Evolution, Old and New,' was not written for the kind of people whom Mr. Romanes calls men of science; if "men of science" means men like Mr. Romanes, I trust they say well who maintain that I am not a man of science; I believe the men to whom Mr. Romanes refers to be men, not of that kind of science which desires to know, but of that kind whose aim is to thrust itself upon the public as actually knowing. 'Evolution, Old and New,' could be of no use to these; certainly, it was not intended as an insult to them, but if they are insulted by it, I do not know that I am sorry, for I value their antipathy and opposition as much as I should dislike their approbation: of one thing, however, I am certain--namely, that before 'Evolution, Old and New,' was written, Professors Huxley and Tyndall, for example, knew very little of the earlier history of Evolution. Professor Huxley, in his article on Evolution in the ninth edition of the 'Encyclopædia Britannica,' published in 1878, says of the two great pioneers of Evolution, that Buffon "contributed nothing to the general doctrine of Evolution,"[379] and that Erasmus Darwin "can hardly be said to have made any real advance on his predecessors."[380]

Professor Haeckel evidently knew little of Erasmus Darwin, and still less, apparently, about Buffon.[381] Professor Tyndall,[382] in 1878, spoke of Evolution as "Darwin's theory"; and I have just read Mr. Grant Allen as saying that Evolutionism "is an almost exclusively English impulse."[383]

Since 'Evolution, Old and New,' was published, I have observed several of the so-called men of science--among them Professor Huxley and Mr. Romanes--airing Buffon; but I never observed any of them do this till within the last three years. I maintain that "men of science" were, and still are, very ignorant concerning the history of Evolution; but, whether they were or were not, I did not write 'Evolution, Old and New,' for them; I wrote for the general public, who have been kind enough to testify their appreciation of it in a sufficiently practical manner.

The way in which Mr. Charles Darwin met 'Evolution, Old and New,' has been so fully dealt with in my book, 'Unconscious Memory;' in the 'Athenæum,' Jan. 31, 1880; the 'St. James's Gazette,' Dec. 8, 1880; and 'Nature,' Feb. 3, 1881, that I need not return to it here, more especially as Mr. Darwin has, by his silence, admitted that he has no defence to make.

I have quoted by no means the moat exceptionable parts of Mr. Romanes' article, and have given them a permanence they would not otherwise attain, inasmuch as nothing can better show the temper of the kind of men who are now--as I said in the body of the foregoing work--clamouring for endowment, and who would step into the Pope's shoes to-morrow if we would only let them.

FOOTNOTES:

[376] See p. 44, and the whole of chap. v., where I say of this supposition, that "nothing could be

conceived more foreign to experience and common sense."

[377] 'Fortnightly Review,' March 1, 1882, pp. 344, 345.

[378] 'Saturday Review,' May 31, 1879, pp. 682-3.

[379] P. 748.

[380] *Ibid.*

[381] See pp. 71-73.

[382] 'Nineteenth Century' for November, pp. 360, 361.

[383] 'Fortnightly Review,' March, 1882.

CHAPTER II.

ROME AND PANTHEISM.

Evolution would after all be a poor doctrine if it did not affect human affairs at every touch and turn. I propose to devote the second chapter of this Appendix to the consideration of an aspect of Evolution which will always interest a very large number of people--the development of the relation that may exist between religion and science.

If the Church of Rome would only develop some

doctrine or, I know not how, provide some means by which men like myself, who cannot pretend to believe in the miraculous element of Christianity, could yet join her as a conservative stronghold, I, for one, should gladly do so. I believe the difference between her faith and that of all who can be called gentlemen to be one of words rather than things. Our practical working ideal is much the same as hers; when we use the word "gentleman" we mean the same thing that the Church of Rome does; so that, if we get down below the words that formulate her teaching, there are few points upon which we should not agree. But, alas! words are often so very important.

How is it possible for myself, for example, to give people to understand that I believe in the doctrine of the Immaculate Conception or in the Lourdes miracles? If the Pope could spare time to think about so insignificant a person, would he wish me to pretend such beliefs or think better of me if I did pretend them? I should be sorry to see him turn suddenly round and deny his own faith, and I am persuaded that, in like manner, he would have me continue to hold my own in peace; nevertheless, the duty of subordinating private judgment to the avoidance of schism is so obvious that, if we could see a practicable way of bridging the gulf between ourselves and Rome, we should be heartily glad to bridge it.

I speak as though the Church of Rome was the only one we can look to. I do not see how it is easy to dispute this. Protestantism has been tried and failed; it has long ceased to grow, but it has by no means ceased to disintegrate. Note the manner in which it is torn asunder by dissensions, and the rancour which these dissensions engender--a rancour which finds its way into the political and social life of Europe, with incalculable damage to the health and well-being of the world. Who can doubt but that there will be a split even in the Church of England ere so many years are over? Protestantism is like one of those drops of glass which tend to split up into minuter and minuter fragments the moment the bond that united them has been removed. It is as though the force of gravity had lost its hold, and a universal power of repulsion taken the place of attraction. This may, perhaps, come about some day in the material as well as in the spiritual and political world, but the spirit of the age is as yet one of aggregation; the spirit of Protestantism is one of disintegration. I maintain, therefore, that it is not likely to be permanent.

All the great powers of Europe have from numberless distinct tribes become first a few kingdoms or dukedoms, then two or three nations, and now homogeneous wholes, so that there is no chance of their further dismemberment through internal discontent; a process which has been going on for so many hundreds of years all over Europe is not likely to be arrested without ample warning. True, during the Roman Empire the world was practically bonded together, yet broke in pieces again; but this, I imagine, was because the bonding was prophetic and superficial rather than genuine. Nature very commonly makes one or two false starts, and misses her aim a time or two before she hits it. She nearly hit it in the time of Alexander the

Great, but this was a short-lived success; in the case of the Roman Empire she succeeded better and for longer together. Where Nature has once or twice hit her mark as near as this she will commonly hit it outright eventually; the disruption of the Roman Empire, therefore, does not militate against the supposition that the normal condition of right-minded people is one which tends towards aggregation, or, in other words, towards compromise and the merging of much of one's own individuality for the sake of union and concerted action.

See, again, how Rome herself, within the limits of Italy, was an aggregation, an aggregation which has now within these last few years come together again after centuries of disruption; all middle-aged men have seen many small countries come together in their own lifetime, while in America a gigantic attempt at disruption has completely failed. Success will, of course, sometimes attend disruption, but on the whole the balance inclines strongly in favour of aggregation and homogeneity; analogy points in the direction of supposing that the great civilized nations of Europe, as they are the coalition of subordinate provinces, so must coalesce themselves also to form a larger, but single empire. Wars will then cease, and surely anything that seems likely to tend towards so desirable an end deserves respectful consideration.

The Church of Rome is essentially a unifier. It is a great thing that nations should have so much in common as the acknowledgment of the same tribunal for the settlement of spiritual and religious questions, and there is no head under which Christendom can unite with as little disturbance as under Rome. Nothing more tends to keep men apart than religious differences; this certainly ought not to be the case, but it no less certainly is, and therefore we should strain many points and subordinate our private judgment to a very considerable extent if called upon to do so. A man, under these circumstances, is right in saying he believes in much that he does not believe in. Nevertheless there are limits to this, and the Church of Rome requires more of us at present than we can by any means bring ourselves into assenting to.

It may be asked, Why have a Church at all? Why not unite in community of negation rather than of assertion? When I wrote 'Evolution, Old and New,' three years ago, I thought, as now, that the only possible Church must be a development of the Church of Rome; and seeing no chance of agreement between avowed free-thinkers, like myself, and Rome (for I believed Rome immovable), I leaned towards absolute negation as the best chance for unity among civilized nations; but even then, I expressed myself as "having a strong feeling as though Professor Mivart's conclusion is true, that 'the material universe is always and everywhere sustained and directed by an infinite cause, for which to us the word mind is the least inadequate and misleading symbol.'"[384]

I had hardly finished 'Evolution, Old and New,' before I began to deal with this question according to my lights, in a series of articles upon God[385] which appeared in the 'Examiner' during the summer of 1879, and I returned to the same matter more than once in 'Unconscious Memory,' my next succeeding work. The articles I intend recasting and rewriting, as they go upon a false assumption; but subsequent reflection has only confirmed me in the general result I arrived at--namely, the omnipresence of mind in the universe.

I have therefore come to see that we can go farther than negation, and in this case--a positive expression of faith as regards an invisible universe of some sort being possible--a Church of some sort is also possible, which shall formulate and express the general convictions as regards man's position in respect of this faith. I think the instinct which has led so many countries towards a double legislative chamber, and ourselves, till at any rate quite recently, to a double system of jurisprudence, law and equity, was not arrived at without having passed through the stages of reason and reflection. There are a variety of delicate, almost intangible, questions which belong rather to conscience than to law, and for which a Church is a fitter tribunal--at any rate for many ages hence--than a parliament or law court. There is room, therefore, for both a State and a Church, each of which should be influenced by the action of the other.

I do not say that I personally should like to see the Church of Rome as at present constituted in the position which I should be glad to see attained by an ideal Church. If it were in that position I would attack it to the utmost of my power; but I have little hesitation in thinking that the world with a very possible feasible Church, would be better than the world with no Church at all; and, if so, I have still less hesitation in concluding, for the reasons already given, that it is to Rome we must turn as the source from which the Church of the future is to be evolved, if it is to come at all.

For the new, if it is to strike deep root and be permanent, must grow out of the old, without too violent a transition. Some violence there will always be, even in the kindliest birth; but the less the better, and a leap greater than the one from Judaism to Christianity is not desirable, even if it were possible. As a free-thinker, therefore, but also as one who wishes to take a practical view of the manner in which things will, and ought to go, I neither expect to see the religions of the world come once for all to an end with the belief in Christianity--which to me is tantamount to saying with Rome--nor am I at all sure that such a consummation is more desirable than likely to come about. The ultimate fight will, I believe, be between Rome and Pantheism; and the sooner the two contending parties can be ranged into their opposite camps by the extinction of all intermediate creeds, the sooner will an issue of some sort be arrived at. This will not happen in our time, but we should work towards it.

When it arrives, what is to happen? Is Pantheism to absorb Rome, and, if so, what sort of a religious formula is to be the result? or is Rome so to modify her dogmas that the Pantheist can join her without doing too much violence to his convictions? We who are outside the Church's pale are in the habit of thinking that she will make little if any advances in

our direction. The dream of a Pantheistic Rome seems so wild as hardly to be entertained seriously; nevertheless I am much mistaken if I do not detect at least one sign as though more were within the bounds of possibility than even the most sanguine of us could have hoped for a few years back. We do not expect the Church to go our whole length; it is the business of some to act as pioneers, but this is the last function a Church should assume. A Church should be as the fly-wheel of a steam-engine, which conserves, regulates and distributes energy, but does not originate it. In all cases it is more moral and safer to be a little behind the age than a little in front of it; a Church, therefore, ought to cling to an old-established belief, even though her leaders know it to be unfounded, so long as any considerable number of her members would be shocked at its abandonment. The question is whether there are any signs as though the Church of Rome thought the time had come when she might properly move a step forward, and I rejoice to think, as I have said above, that at any rate one such sign--and a very important one--has come under my notice.

In his Encyclical of August 4, 1879, the Pope desires the Bishops and Clergy to restore the golden wisdom of St. Thomas Aquinas, and to spread it far and wide. "Vos omnes," he writes, "Venerabiles Fratres, quam enixe hortamur ut ad Catholicæ fidei tutelam et decus, ad societatis bonum, ad scientiarum omnium incrementum auream Sancti Thomæ sapientiam restituatis, et quam latissime propagetis." He proceeds then with the following remarkable passage: "We say the wisdom of St.

Thomas. For whatever has been worked out with too much subtleness by the doctors of the schools, or handed down inconsiderately, whatever is not consistent with the teachings of a later age, or finally, is in any way NOT PROBABLE, We in no wise intend to propose for acceptance in these days."[386]

It would be almost possible to suppose that these words had been written inadvertently, so the Pope practically repeats them thus: "We willingly and gratefully declare that whatsoever can be excepted with advantage, is to be excepted, no matter by whom it has been invented."[387]

The passage just quoted is so pregnant that a few words of comment may be very well excused. In the first place, I cannot but admire the latitude which the Pope not only tolerates, but enjoins: he defines nothing, but declares point blank that if we find anything in St. Thomas Aquinas "not consistent with the assured teachings of a later age, or finally IN ANY WAY NOT PROBABLE"--(what is not involved here?)--we are "in no wise to suppose" that it is being proposed for our acceptance. But it is a small step from allowing latitude in accepting or rejecting the parts of St. Thomas Aquinas which conflict with the assured result of later discoveries to allowing a similar latitude in respect, we will say, of St. Jude; and if of St. Jude, then of St. James the Less; and if of St. James the Less, then surely ere very long of St. James the Greater and St. John and St. Paul; nor will the matter stop there. How marvellously closely are the two extremes of doctrine approaching to one another! We, on the one hand, who begin with *tabulæ rasæ* having made a clean sweep of every shred of doctrine, lay hold of the first thing we can grasp with any firmness, and work back from it. We grope our way to evolution; through this to purposive evolution; through this to the omnipresence of mind and design throughout the universe; what is this but God? So that we can say with absolute freedom from *équivoque* that we are what we are through the will of God. The theologian, on the other hand, starts with God, and finds himself driven through this to evolution, as surely as we found ourselves driven through evolution to the omnipresence of God.

Let us look a little more closely at the ground which the Church of Rome and the Evolutionist hold in common. St. Paul speaks of there being "one body and one spirit," and of one God as being "above all, and through all, and in you all."[388] Again, he tells us that we are members of God's body, "of his flesh and of his bones;"[389] in another place he writes that God has reconciled us to himself, "in the body of his flesh,"[390] and in yet another of the Spirit of God "dwelling in us."[391] St. Paul indeed is continually using language which implies the closest physical as well as spiritual union between God and those at any rate of mankind who were Christians. Then he speaks of our "being builded together for an habitation of God through the spirit,"[392] and of our being "filled with the fulness of God."[393] He calls Christian men's bodies "temples of the Holy Spirit,"[394] in fact it is not too much to say that he regarded Christian men's limbs as the actual living organs of God himself, for the expressions quoted above--and many others could be given--come to no less than this. It follows that since any man could unite himself to "the flesh and bones" of God by becoming a Christian, Paul had a perception of the unity at any rate of human life; and what Paul admitted I am persuaded the Church of Rome will not deny.

Granted that Paul's notion of the unity of all mankind in one spirit animating, or potentially animating the whole was mystical, I submit that the main difference between him and the Evolutionist is that the first uses certain expressions more or less prophetically, and without perhaps a full perception of their import; while the second uses the same expressions literally, and with the ordinary signification attached to the words that compose them. It is not so much that we do not hold what Paul held, but that we hold it with the greater definiteness and comprehension which modern discovery has rendered possible. We not only accept his words, but we extend them, and not only accept them as articles of faith to be taken on the word of others, but as so profoundly entering into our views of the world around us that that world loses the greater part of its significance if we may not take such sayings as that "we are God's flesh and his bones" as meaning neither more nor less than what appears upon the face of them. We believe that what we call our life is part of the universal life of the Deity--which is literally and truly made manifest to us in flesh that can be seen and handled--ever changing, but the same yesterday, and to-day, and for ever.

So much for the closeness with which we have come together on matters of fact, and now for the *rapprochement* between us in respect of how much conformity is required for the sake of avoiding schism. We find ourselves driven through considerations of great obviousness and simplicity to the conclusion that a man both may and should keep no small part of his opinions to himself, if they are too widely different from those of other people for the sake of union and the strength gained by concerted action; and we also find the Pope declaring of one of the brightest saints and luminaries of the Church that we need not follow him when it is plainly impossible for us to do so. Is it so very much to hope that ere many years are over the approximation will become closer still?

I have sometimes imagined that the doctrine of Papal Infallibility may be the beginning of a way out of the difficulty, and that its promoters were so eager for it, rather for the facilities it afforded for the repealing of old dogmas than for the imposition of new ones. The Pope cannot, even now, under any circumstances, declare a dogma of the Church to be obsolete or untrue, but I should imagine he can, in council, *ex cathedra*, modify the interpretation to be put upon any dogma, if he should find the interpretation commonly received to be prejudicial to the good of the Church: and if so, the manner in which Rome can put herself more in harmony with the spirit of recent discoveries, without putting herself in an illogical position, is not likely to escape eyes so keen as those of the Catholic hierarchy. No sensible man will hesitate to admit that many an interpretation which was natural to and suitable for one age is unnatural to and unsuitable for another; as circumstances are always changing, so men's moods and the meanings they attach to words, and the state of their knowledge changes; and hence, also, the interpretation of the dogmas in which their conclusions are summarized. There is nothing to be ashamed of or that needs explaining away in this; nothing can remain changeless under changed conditions; and that institution is most likely to be permanent which contains provision for such changes as time may prove to be expedient, with the least disturbance. I can see nothing, therefore, illogical or that needs concealment in the fact of an infallible Pope putting a widely different interpretation upon a dogma now, to what a no less infallible Pope put upon the same dogma fifteen hundred, or even fifteen years ago; it is only right, reasonable, and natural that this should be so. The Church of England may have made no provision for the virtual pruning off of dogmas that have become rudimentary, but the Encyclical from which I have just quoted leads me to think that the Church of Rome has found one, and, in her own cautious way, is proceeding to make use of it. If so, she may possibly in the end get rid of Protestantism by putting herself more in harmony with the spirit of the age than Protestantism can do. In this case, the spiritual reunion of Christendom under Rome ceases to be impossible, or even, I should think improbable. I heartily wish that my conjecture concerning future possibilities is not unfounded.

Scientists have been right in preaching evolution, but they have preached it in such a way as to make it almost as much of a stumbling-block as of an assistance. For though the fact that animals and plants are descended from a common stock is accepted by the greater and more reasonable part of mankind, these same people feel that the evidence in favour of design in the universe is no less strong than that in favour of evolution, and our scientists, for the most part, uphold a theory of evolution of which the cardinal doctrine is that design and evolution have nothing to do with one another; the jar they raise, therefore, is as bad as the jar they have allayed.

It has been the object of the foregoing work to show that those who take this line are wrong, and that evolution not only tolerates design, but cannot get on without it. The unscrupulousness with which I have been attacked, together with the support given me by the general public, are sufficient proofs that I have not written in vain.

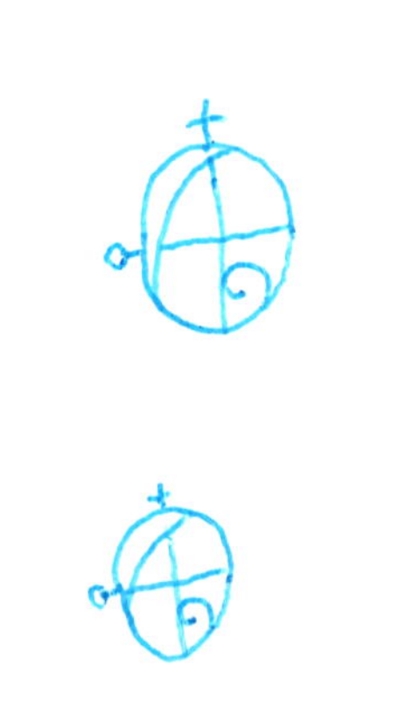

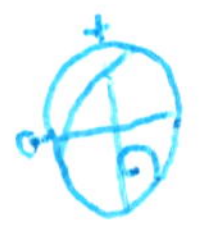

Made in the USA
Las Vegas, NV
07 May 2021

22611511R00243